*Life beyond Molecules and Genes*

# Life beyond Molecules and Genes

## HOW OUR ADAPTATIONS MAKE US ALIVE

Stephen Rothman

TEMPLETON PRESS

Templeton Press
300 Conshohocken State Road, Suite 550
West Conshohocken, PA 19428
www.templetonpress.org

Designed and typeset by Gopa and Ted2, Inc.

Library of Congress Cataloging-in-Publication Data

Rothman, Stephen.
Life beyond molecules and genes : how our adaptations
make us alive /
Stephen Rothman.
p. cm.
Includes bibliographical references and index.
ISBN-13: 978-1-59947-250-8 (pbk. : alk. paper)
ISBN-10: 1-59947-250-3 (pbk. : alk. paper)
1. Adaptation (Biology)  2. Life.
3. Evolution (Biology)  I. Title.
QH546.R68 2009
578.4—dc22
2009006181

Printed in the United States of America
09 10 11 12 13 14    10 9 8 7 6 5 4 3 2 1

For Alex, Bennett, Elijah, Jack,
Jessi, and Tyrone

*The world is too much with us; late and soon,*

*Getting and spending we lay waste our powers:*

*Little we see in nature that is ours;*

*We have given our hearts away, a sordid boon!*

WILLIAM WORDSWORTH,
"THE WORLD IS TOO MUCH
WITH US," 1806

———— ⚬ ————

# Contents

# Preface

LIFE BEYOND MOLECULES AND GENES is not a "how to" book about the good life and how to live it. Nor is it about the human psyche—about pleasure, fulfillment, love, conflict, pain, and struggle. Nor is it an esoteric philosophical treatise about "being," "materialism," and the limits of human understanding. Neither is it a book about deep religious or theological questions such as what constitutes a moral life or how to explain life's essence through a belief in and revelation of God. Nor is it a historical analysis or compendium comparing and contrasting these and other ideas about life's nature down through the ages to modern times.

The essence of this book is biological. It presents an extended argument for a particular point of view from biology—and from science more broadly—about what it means to say that something is alive. This does not mean that it has nothing to do with these other subjects, with how we live life, with our psychology, philosophy, and religion, or more broadly with humankind's search to understand life's nature. For me, at least, it is hard to read without thinking about them. This is in no small measure because the question it poses—what makes certain material bodies alive?—is the most profound that science can ask about living things. But it is also because in seeking an answer, *Life beyond Molecules and Genes* concludes that a purely material explanation for life will just not do. Life, it argues, is not to be found in any particular physical or chemical embodiment, but in certain contingent, transcendent, and even immaterial activities.

For this reason, we can say that *Life beyond Molecules and Genes* has a "spiritual" as well as a scientific dimension. Indeed, any scientific inquiry into the deepest nature of life cannot help but raise equally deep questions about life's existential character and circumstances. As its contents make clear, the deeper we look, the more transparent it becomes that at the most fundamental level, the science of life and our religious and philosophical beliefs, as well as the mundane attitudes and ways of everyday life, are knotted. Indeed, how could it be otherwise?

This said, *Life beyond Molecules and Genes* comes to its particular point of view in a novel way for modern biology. It combines a traditional "physiological" perspective—an examination of the basis of the internal activities of our bodies—with evolutionary theory. It is in doing this that it exposes the truth, both obvious and shrouded, that life is not to be found in genes or molecules, in chemical reactions, in what we are made of more broadly, or even how we are put together, but in what we do and why we do it. In this I include everything from the beating of cilia, to the beating of our heart, to our deepest psychological drives and broadest social behaviors. As such, *Life beyond Molecules and Genes* provides a much-needed response to molecular biology and genetic determinism gone amok, to the current infatuation with DNA and the seemingly never-ending claims that it and its molecular cohorts explain everything worth explaining about life—from the depraved, to the commonplace, to the exalted.

I imagine that those who will find *Life beyond Molecules and Genes* interesting are people whose inquisitiveness leaves them uncertain about both religious and materialistic claims about life, who harbor unanswered questions and troublesome doubts, rather than certainties, in both realms. They are not so much agnostics, as skeptics, doubters of others' certainties. Yes, they wonder about the existence of God, of heaven and hell, but they also wonder about the striking dissonance between the matter—molecules and genes—that science tells us explains life and the life we live, much less the life we lead.

Of course the incurious, those who don't give a fig about what they are or their place in nature, are not likely to be interested in what follows. But it also would not be surprising if those who care deeply about such questions but who hold strong and impermeable beliefs—some devoutly religious people who place man's deepest essence outside of scientific discourse, or the devoutly *irreligious* who hold to an unshakable materialism in which all that matters is matter—conclude, in all likelihood without reading what it has to say, that this book is not only of no interest, but defamatory. While it is certainly possible to pique the interest of the initially disinterested, it seems fruitless and foolhardy to believe that one can win over people for whom the truth is already known and incontrovertible. I can only say that *Life beyond Molecules and Genes* offers a clarification that helps explain what seems to be the ever-widening rift between religion and materialist science, and in so doing, exposes a deeper harmony in their common quest to understand who and what we are.

There are three additional points that I would like to make. First, the contents of this book are couched in terms of evolutionary and gene theory. From a modern perspective, evolutionary theory refers to life's observable mechanisms, processes, shapes, and species having come into being as the result of the actions of natural selection on various mutations, modifications to our being exemplified by structural changes to the DNA molecule, or on the consequences of these changes, while gene theory claims that life arises from our genes, understood as the DNA code for proteins.

Even though the book's subject matter, if not its conclusions, stands on its own and does not depend on either evolutionary or gene theory, it is discussed in this context because the seminal question it asks is posed in light of them. Not only that, but considered apart from them, its message, meaning, and significance would be obscure, if not completely unintelligible, to a modern scientific sensibility, floating, as it were, outside its conception of the world. What is more, as we shall see, though we can talk about being alive without reference to biological evolution, the phenomenon cannot be understood without taking it into account.

This said, however reasonable, however much evidence there may be, we should be clear that neither the mechanisms of evolution nor, outside of the genetic code, how DNA genes embody our various aspects have been established as a matter of scientific law. That is why we talk of evolutionary and genetic theories, not laws. As for evolution, thanks to the fossil record we know a great deal about the evolution of species, and though time is natural selection's ally, its security blanket, the obverse is that it seems far-fetched to think that we can know what selective events took place in the long distant past—over evolution's lengthy four-billion-year history—with any sort of detail and rigor. However perceptive and thoughtful our conjectures may be, that is all they are, and probably all they can ever be. As for genetic theory, as particular (molecular) as our concept of the gene has become, beyond the encoded proteins themselves it remains to be shown how DNA's genetic code gives rise, directly and inevitably, to our various complex incarnations and processes. Some of my personal concerns and questions about these issues surface, mostly as sidebars, in *Life beyond Molecules and Genes*.

The second point is that it is important to distinguish the idea that life is inherent to, one and the same as its material embodiment, its genes and molecules and all they are and do—what we can call "biological materialism"—from the larger philosophical system of materialism. Though materialist ideas are ancient and manifold, materialists today commonly accommodate their views to our current scientific understanding of the physical world. As such, they usually accede to energy being a form of matter, perhaps its most fundamental form, and include space, its curvature, the uncertainties of quantum mechanics and field theory, and even time in their conception of the material world.

But perhaps even more important here, the modern materialist not only believes that all objects are comprised of matter (whatever the term connotes and encompasses), but that *all events in the physical universe arise from or are the result of interactions of one sort or another between material objects*. In reading *Life beyond Molecules and Genes,* it is essential to keep in mind that the materialism it questions is limited to *biological materialism,* to life being inherent to a particular material body, not this broader system of materialist belief, with which its contents are entirely compatible.

However, compatibility is not affirmation. *Life beyond Molecules and Genes* does not affirm materialist views of reality to the exclusion of various forms of idealist thought, which consider among other things ideas as a necessary element in any description of reality. Nor does it choose between monist and dualist visions of the world. It is by intention resolutely neutral on these complex philosophical subjects. It presents, as said, a biological thesis.

Biological materialism's kissing cousin is "biological reductionism." Like biological materialism, it says that life can be completely understood in terms of, or can be reduced to, its underlying material components. And so, as *Life beyond Molecules and Genes* criticizes biological materialism, it also criticizes this reductionist philosophy. In a companion volume, *Lessons from the Living Cell: The Limits of Reductionism* (New York: McGraw-Hill, 2002), I evaluate this approach to the study of life as scientific practice. Methodological reductionism, or what I call "strong micro-reductionism," claims that a complete understanding (strong) of life can be obtained from the study of its most intimate properties (micro-). That is, it is the view that elucidating the underlying molecular and other microscopic physical, chemical, and anatomical properties of living things can secure a complete understanding of their broader nature.

Three, *aliveness* is a single observable phenomenon however varied its incarnations, and as such can have but one definition. It can only allow multiple definitions if they have identical meanings or otherwise logically commute. There cannot be different, nonidentical, noncommuting, but equally legitimate definitions of life—for example, of life seen as molecules, or their reactions, or in genes, or their products, or as cells, ecology, or complexity, and so on, as is sometimes suggested. However helpful, it is simply not permissible.

Finally, *Life beyond Molecules and Genes* discusses scientific ideas that may at times seem arcane. Yet, what it has to say is straightforward, even simple—indeed, perhaps shockingly so. I have tried to use nontechnical terms and everyday language where possible, and hope that in doing so I have had at least some success in making its ideas clear and accessible to any interested reader with a basic education in biology and science.

# Acknowledgments

I WOULD LIKE to express special gratitude to my dear friend Aaron Lukton for his willingness to read the manuscript in its early incarnations and for our many wonderful discussions about evolution and biology over curry soba at Kirala. I would also like to thank Amy Bianco who encouraged me early on to examine the question of life's essential nature through the lens of a physiologist. As always, I could not have persevered in this task without the continued love and encouragement of my wife, Doreen. And finally, I would like to express deep appreciation to those at Templeton Press for their interest in the book, to Natalie Lyons Silver for husbanding the manuscript, to Amy Wagner for editing it, and to all the others at Templeton who made its production possible.

*Life beyond Molecules and Genes*

# What Is It That Makes an Object Alive?

And God formed man of the dust of the ground, and breathed into his nostrils
the breath of life; and man became a living soul.
**—GENESIS 2:7 (KING JAMES TRANSLATION)**

When a thing has existence alone
it is mere dead weight
Only when it has *wu* (nothingness, emptiness,
non-existence), does it have life.
**—LAO TZU, *TAO TE CHING* (TRANSLATED BY JONATHAN STAR)**

BEING HUMAN is to deny certain realities of our circumstances,
such truths that we find too painful to contemplate or unpleas-
ant facts of our lives that seem intractable to us. Throughout the
twentieth century, the discipline of biology was in such a state of
denial. It determined to deny the meaningfulness of nothing less
than its central question: what is it that makes something living? In fact, it reasoned that *only* by ignoring it could progress
be made.

It was this grand question that in the early nineteenth century
justified the need for a discipline focused exclusively on deciphering the nature of living things, and nothing less than understanding the pivotal fact of life—that living things are comprised of
small objects called cells, which are alive themselves—came from
posing it. However, in the modern era, it has been thought unan-
swerable, and accordingly seen as an impediment to progress to
be diligently avoided. Unlike physical scientists, biologists came

to see attempts to answer the essential question of their discipline as counterproductive. They thought that science would be better served if they learned as much as possible about the intimate chemical and physical character of living things, what they are comprised of and how they work, rather than speculating about their ultimate nature. After all, wasn't this really what we wanted to know? It was hoped that this practical approach would be more fruitful and more instructive, even if we remained forever unable to say what made something living.

Given what has happened, it is hard to question this judgment. The twentieth century was a period of unprecedented progress in understanding life's nature. After millennia in the dark, in a matter of a mere one hundred or so years, we learned with astonishing clarity and detail what substances and structures comprise living things and how they carry out their major activities. But a curious thing happened. Given the determination of biology to relegate its great question to the dustbin of history as pointless or hopeless, it just could not be avoided. In an astounding paradox, the shunned question found an answer. In an unexpected contradiction, avoidance led to understanding, and many scientists today believe that we have actually fathomed what beyond cells makes something living.

In this view, science has mastered, or at least has come close to mastering, the phenomenon of life with the same rigor as the rest of the physical world. All that life is—and this is not to diminish it—is a particular, though remarkably complex, articulation of physical and chemical law. At the center is the giant deoxyribonucleic acid molecule (DNA) and its genes that not only explain how life issues from parent to offspring and how life's various features are inherited, but explains how those features come into being, or become real. As a result, uncovering the role of DNA was a scientific achievement equal to the greatest in physics and chemistry. Moreover, the remarkable properties of this molecule, and the protein molecules for which it holds the code, were reassuringly found to be explicable in terms of the selfsame physics and chemistry.

These enormous advances were in great part attributable to the philosophical belief known as *materialism*. When observers of modern life say that we have become too materialistic, they are usually referring to our obsession with material possessions. But the materialism of contemporary life is far more than cars and television sets. So much that has happened during the past two hundred years has made us view our existence in less transcendent and increasingly materialistic terms.

This includes momentous events such as the rise of societies that are not tied to religious belief and the accomplishments of industrialization, medicine, and agriculture that have greatly improved the lives of so many people. And whether reflecting on Marx's dialectical materialism or notions of the death of God raised by the *Shoah*, many have become convinced that life, like the world around us, is merely a materialistic matter.

Of all the materialistic predilections of modern life, scientific and technological materialism is second to none. We expend great resources and intellect, through engineering and chemistry, to invent new machines and new molecules. Relativity theory notwithstanding, whether as bosons, quarks, or various solid, liquid, and gaseous states of matter, we seek to understand the physical world by grasping its material nature. And the science of life is no exception. Our gaze has increasingly focused on our material ingredients, on the molecules of which we are made, especially the master DNA molecule.

And so, it has come to pass that many people, including some of our most accomplished intellectuals and scientists, believe that life in all its forms is wholly attributable to the matter of which it is comprised. There is nothing else. This sad truth of biology is usually taught by example. In a long litany, we are exposed to one, another, and yet another underlying physical or chemical incarnation of some aspect of life. We learn what genes are and how they work, about diseases caused by defects in them and their product proteins, about the genetic basis of development, about how our brain works with its neurons and synapses, about the molecular basis and chemical treatment of disease—from metabolic disorders such as diabetes to mental ones such as schizophrenia—and finally we are taught about all manner and means of biological molecules—from nucleic acids to enzymes to hormones to metabolites to neurotransmitters to antibodies— that affect, sustain, protect, and indeed comprise us.

In this view, life is coextensive with, or one and the same thing as, a particular physical and chemical system. They are identical—two aspects of the same reality. To understand this system is to understand life. What is more, most biologists today would probably agree that we are able to describe that system quite well, if not in all its particulars, and as such understand life. DNA and its genes give rise to proteins, and the actions and reactions of proteins performed in concert and organized in space and over time give rise to life in both cell and organism. This is what makes particular material objects alive. It is life, as modern science understands it.

This understanding is not only simple and elegant, it is absolutely breathtaking. Yet, with all due appreciation, is it true? Is it true that understanding our physical and chemical manifestation is the same as understanding what makes us alive? If it is, then we have to ask how this comes about. Not how DNA makes proteins, or how proteins make structures or allow reactions, but how our material embodiment makes us alive? If science seeks to be explanatory, as it must if it is to bear its name, it should be able to tell us how this transformation occurs. It should be able to specify with all necessary precision and detail the bridge that is crossed as we move from the material to the living realm. And yet I doubt that any of you have ever been offered such an explanation in a biology class or anywhere else. In over forty years of teaching physiology, I never provided one, nor did any of my colleagues. The reason is simple—it does not exist. We have no explanation.

This does not concern many, perhaps most, biologists. They believe it is explanation enough *to claim* an identity between life and its material incarnation, because *ipso facto*, by their very nature, they are one in the same thing. But however emboldening, comforting, or reassuring, this is not a statement of science. Without elucidation, without explication, without proof, simply claiming an identity between two things can never be more than an assumption, a supposition, an expectation, a guess, a statement of authority, not science, however confidently declared.

Scientists who appreciate this truth often deal with the difficulty in one of two ways. Either they see life—aliveness—as a vague and indefinable concept that we cannot hope to stipulate in a rigorous fashion, or to the contrary they are convinced that science will do just that at some future time. For obvious reasons, we can call the former group pessimists, and the latter group optimists.

There is not much to say about the pessimists' viewpoint, other than that they believe it is warranted. The optimists most commonly express their confidence through their belief in genetics and chemistry. Through them, everything will (must) be revealed sooner or later, in one way or another. They are persuaded that whatever the deficiencies in our current understanding, however the science of life unfolds (and admittedly they don't know exactly how), eventually, when we know enough about our genes, enough about our chemistry, the sought-after explanation will become self-evident, rising like Botticelli's *Venus* from the sea.

But the truth is that whether despairing or brimming with confidence, what both pessimists and optimists profess is really no different than asserting an unproven identity as fact. However heartfelt, their sensibilities are more devout than rational. As will be explained in what follows, as a matter of reason, it has been historically and remains demonstrably true that the state we refer to as "being alive" *cannot* be understood in terms of the material composition of organisms, however broadly we cast our net. The simple fact is that however instructive and indeed however indispensable information about life's material nature is to understanding living things—and it is both—when examined from this perspective alone, life's essence is lost.

On the other hand, if we claim that life's identity with its material incarnation is incorrect, then we must be able to conceptualize a satisfactory alternative. I use the word "must" because if we cannot provide one then the claim, fallacious though it may be, can be made that the identity of life's material incarnation with life itself is assured by exclusion however ignorant we may be about how one leads to the other. Put simply, there is no other choice.

In this light, *Life beyond Molecules and Genes* offers two propositions, the first being the predicate for the second. First, the phenomenon of life is *not* coextensive with or identical to its material incarnation. It is not to be found in the material instantiation of living things, not in any or all of their necessary materiality, not in DNA or proteins, not in both together, not in the reactions and interactions of life's molecules more broadly, not in the structures they form and in which they are located, nor in how they are organized, nor finally, in any of their associated physical properties.

The second proposition declares that an alternative account of the phenomenon is not only possible, but can be specified and, more than that, fulfills the two central requirements of any such representation. It *defines* life clearly and provides a means to *distinguish* living things from all other natural objects. It is important to understand the significance of this claim. I am aware of no other explanation for life that fulfills these two essential requirements. What is proposed here does, and as such is unique. But more than that, if it is reasonable as well as unique, then, as explained in the preface, any other explanation that we might come up with must be commensurate or correspond to it in all respects. It must, in essence, be identical. As such, it sets a valuable standard against which we can judge other explanations.

*Life beyond Molecules and Genes*'s conception of life is based on a reconsideration of the relationship between the *material nature of living things* and their *evolution*. It is not so much a new concept, as a new perspective. It casts well-known and well-rehearsed facts in a new light. It is new in the same way that a Gestalt shift of a visual illusion furnishes a new image.

It has to do with *biological adaptations.* An adaptation is any feature of an organism, including its psychological, behavioral, and social aspects that helps it cope with its environmental circumstances. As Darwin understood adaptations, they are the central and, in fact, the only meaningful product of evolution. For all intents and purposes, biological evolution is the evolution of adaptations.

The remarkable variety of these phenomena can be sampled on the pages of any one of many good books published over the years about Darwin and the theory of evolution, including numerous textbooks, but most wonderfully in Darwin's own acute descriptions in *On the Origin of Species* itself. Though exceptionally diverse in nature, as usually described, adaptive phenomena have one thing in common—they pertain to life as it is lived from the outside, or perhaps more accurately, as it is seen from the outside. From this external perspective, adaptations are understood to be the skills necessary to obtain food, to defend against becoming food for others, to endure extremes in temperature and weather more generally, and critically, to procure sex—all in the service of survival and propagation.

As such, we say that adaptations are properties of living things. They are what living things do to secure their continued existence. Given this understanding, we can say, and it seems self-evidently true, that our adaptive properties exist because we are alive. And if adaptations are properties of living things, then our aliveness must be due to something else, something that allows adaptations to exist. Indeed, it is our unmistakable perception that our aliveness is intrinsic, built-in, that we stand on our own as living beings distinct from the world in which we live and with which we interact. There appear to be two kinds of life traits—those that impart life, and those, like adaptations, that are features of otherwise living objects.

This brings us to the crux of the matter, to the aforementioned Gestalt shift. It is the contention of this book that this statement is not only false, but that it has it backward. However counter-intuitive, our adaptations do not exist because we are alive, we are alive because they exist. The claim is that they and they alone carry the elusive property of aliveness.

To say that we are alive because of our adaptations is to declare something else of importance. It is to claim that life is not inherent, that there is no life, no living object, as mere inherence. As we shall see, adaptations do not exist in and of themselves, as isolated phenomena. They are the product of interactions of one sort or another with the environment.

This is not to say that the material features that *give rise* to adaptations do not exist in the absence of these interactions—they certainly do—but the adaptations themselves, the products of the interactions, do not. And if it is our adaptations that make us alive and they arise out of interactions with the environment, then it follows that there is no feature or group of features that bestow life on an object *free of this essential interaction*. Only in this way do certain material objects gain and sustain life. Only then, only together, object and environment do they signify our quintessence.

This is all well and good, but why then do we seem to be alive inherently? Why is it our clear and unambiguous sense that our adaptive characteristics, whether predatory or defensive, are properties of otherwise living things? Have we been tricked, and if so, by whom or what? As it happens, we have been tricked. Our sense that our aliveness is inherent is due to a lack of awareness of something else, something almost completely hidden from view. And it is a big something. It is the remarkable world of the adaptations of life as it is lived from the *inside—the internal adaptations of life*—the adaptations of the body's inner workings. As we shall see, in realizing that life's inner workings are also adaptive, a new, or at least newly reimagined, view of life emerges. This new view includes the realization that the evolution of the diverse forms and functions of life's adaptations are not merely the result of chance occurrences as so often thought, but are the predictable consequence of the uncompromising demands of nature's abiding physical, chemical, and mathematical laws.

Before telling this story, it is important to point out that though what I have to say is grounded in modern biology with its molecules and genes, it owes its principal debt to Charles Darwin and his theory of evolution by natural selection, the only biological theory that is independent of any particular chemistry and physics. Its inspiration flows directly out of the transformational ideas about the origin of the species that were first fully articulated independently by Charles Darwin and Alfred Russel Wallace, as well as related, through less well-known sentiments about life's nature by the eminent nineteenth-century French physiologist Claude Bernard.

# CHAPTER I

# Life Forces and Vital Substances
## The Path from Vitalism to Materialism

———————————— ⮂ ————————————

Young men called by their genius to the improvement of science cannot
too strongly safeguard themselves against this sort of error [invoking vital
forces to account for the properties of life]. They must early become
accustomed to learn to say to themselves I don't know.
—**FRANÇOIS MAGENDIE, PRECIS, 1833**

SOME ENDED UP concluding that the search was a fool's game,
and that attempts at exegesis were futile, ultimately and inevitably doomed to failure. Life's fundamental nature seemed beyond
human comprehension, beyond our ability to understand, so different from the rest of the material world, even more remarkable
and unexplainable. Even the great Pasteur said in resignation
that it could not be defined.

Either life's essence was miraculous, the product of an intelligence outside our imagining, or if it was the consequence of ordinary processes, they were so complex, so obscure, so hidden that
we could never decipher them. However inexplicable our physical environment was with its endless oceans and lifeless deserts;
towering and unscalable mountains; ever-changing, terrifying,
and life-threatening weather; with earthquakes, volcanoes, and
all kinds of other cataclysms; and with the wondrous and awesome firmament, the astonishing character and variety of life was
second to no other mystery of nature.

## Looking beyond the Surface

What was life? How did it differ from the rest of the material world? How did it come into being? Where did it come from? Could we ever hope to explain such mysteries? And yet despite this daunting prospect, humans have an easy ability to distinguish most living things from inanimate objects. Our brain analyzes what we see—form, color, motion, growth—what we feel, hear, smell, and taste, and we grasp its meaning. Is what we are facing living or inanimate? And this ability is not unique to humans; it is a critical and distinguishing feature of life in multitudes of animal species, even those without conscious minds. Their survival depends on it.

In its service, evolution has contrived a rich and varied collection of mechanisms by which organisms sample, assess, and react to their environment, what biologists used to call "irritability," itself an important signifier of life. Above all else, animals must be able to determine what is food and what is not, what may attack it, what cannot, and act accordingly.

Yet however well developed this faculty may be, it is one thing to know that something is alive, and quite another to know why—what makes or causes it to be alive. The first is a utilitarian competence that provides no understanding beyond the elements of the distinction itself. It is like learning to use the remote control for a television set. Having mastered it, you can watch television to your heart's content, turn it on and off, change the channel and the volume, and use whatever other features it may have, without the foggiest notion of how either it or the television set works.

Your ignorance of the machine's inner workings is no impediment. But however well you can use your remote, this skill will never enlighten you about its or the television's underlying causes. In this case, doing and understanding do not commute. So perhaps it is surprising that unlike the remote device, the practical competencies that allow humans to distinguish the animate from

the inanimate *have* been critical to our search for a deeper understanding of the nature of life.

If you think about it for a moment, you will realize why. How
else could we proceed? Where else could we start if not with
what we see, hear, feel, and smell, with our sensations of the
world? Only with this knowledge, however inadequate or superficial, can we wonder about unseen causes. Indeed, making this
transition from superficial identification to speculating about
underlying causes is one of the most remarkable, if not *the* most
remarkable, feature of our species. Perhaps more than anything
else, it sets us apart from all other material objects.

## The Three Traditional Views of Life's Causes

Though our history—cultural, political, philosophical, religious,
and scientific—is filled with all sorts of imaginings about life's
nature, the beginnings of modern scientific thought on the subject can be traced to the labors of a few ancient Greek thinkers:

To Democritus, who imagined atoms beneath the surface of
material objects, we credit reductionism.

To Aristotle we owe the idea that life's underlying cause lies in
its form—its geometry, its symmetry, its essential organization.

And to Galen, the great physician and first experimental biologist, we are indebted for the belief that to understand animal
movement is to understand life.

Between the fall of the Aristotelian system and the emergence
of the modern scientific consensus, roughly from the sixteenth to
the twentieth century, there seemed to be almost as many theories about life's nature as individuals who posed them. Yet people have only imagined objects being *enlivened*, brought to life,
in three ways.

In the first, life is sparked by an entity that enters a particular
body *from the outside*. Its addition gives life; its removal ends it.
Causation is external. Of course, this way of imagining life is a
portrayal of the transcendent soul of our religious metaphysics.
But the idea that life is added from the outside was not solely a

religious belief. To some scholars what was added was natural, not supernatural; material, not immaterial; and though unseen, a distinct and perfectly real aspect of the physical world. For example, for animals and plants respectively, we might think of oxygen and sunlight as such things.

With Aristotle, the second view seeks life's cause in its organization—life-as-organization. This has been imagined in two different ways. Either something within the putative living thing—either there in and of itself, or added to it from outside—gives life to certain matter by causing it to be organized in a particular fashion. Or alternatively, and rather popular today, the lifegiving property is the *product*, not the cause of the organization. Organization itself is lifegiving. Embryonic development can be understood either way.

The final way of thinking about life's causes is the most directly and obviously materialistic. According to this view, life is found in the material substances that comprise animate objects with their attendant physical and chemical properties. We say that life is *immanent* to the object, within and directly attributable to its material contents or at least certain aspects of them, and them alone. There is nothing more, and nothing more need be imagined. Life does not come from the outside, nor does it emerge from a particular organization of things. It is simply and immediately an inherent property of the object's material embodiment.

## The Life Force and Vital Substance

Whatever gives rise to life, whether it comes from the outside, is an organizing principle, or is inherent to the object's corporeality, what exactly is it? Today we can all but specify our substance, what we are made of. But before the modern era, indeed for most of recorded human history, this was impossible. Scholars referred to a vague *life force*, or a *vital substance*, a *materia vitae*, a fundamental material that animated the inanimate.

But no matter what it was called—and it was called many

things—whatever the term and however much descriptive language attended its use, it was never much more than an appellation. The fact was that most notions about the nature of life's substance or about the life force were not merely vague, they were totally vaporous.

The reality was that whatever its nature—vital force or vital substance—it was inaccessible, unfathomable. We could barely describe its attributes, much less its fundamental nature. Did it have mass, embody energy, or was calling it a substance or force merely a metaphor for something else, something different from or beyond substance and energy?

And if it was, what was that something else? Some argued that life was an irreducible phenomenon of nature, and after Newton, likened it to gravity. Despite the mathematical precision of Newton's formula for gravitation, how its force was generated was a mystery. Maybe the deep or intimate nature of life was similar—inscrutable, forever undiscoverable by humans.

## The Great Debate

As is often the case in such situations, a lack of understanding does not mean a lack of explanations. There were many and varied accounts of lifegiving forces and substances and they all suffered from the same deficiency—a lack of, well, force and substance. It was not until the early nineteenth century that it became possible to put meat on the bare bones of the designated patronymic.

Three advances were critical:

- The first was Newton's laws, established more than a hundred years earlier, that described with mathematical precision certain physical properties of material objects and, most important, the cause of their movement.
- The second began and culminated in the work of Antoine Lavoisier in the eighteenth century, and fortunately for us, did so before a revolutionary tribunal decided to end his life by guillotine in 1791, having rejected a plea for mercy with the statement that "the revolution has no need of scientists."

Scientist or not, it should be noted that Lavoisier supported himself as a tax collector for the king.

Along with the British Unitarian minister Joseph Priestley, Lavoisier discovered oxygen (Priestley called it "dephlogisticated air," signifying the *absence* of a substance, phlogiston) and its role in combustion. This was a monumental unearthing for many reasons. In the first place, it was the first time that biology was firmly coupled to chemistry. It was now understood that a chemical substance, oxygen, was required to sustain animal life.

Equally important, Lavoisier's quantitative approach to chemistry foreshadowed the essence of the modern science —he balanced the amount of product produced with the amount of substrate added. Finally, his work gave rise to and provided proof for the notion of elements and compounds. These are immense contributions and made Lavoisier the undisputed father of modern chemistry.

▸ The third advance was cell theory: all living things are made of microscopic, living objects called cells. Cell theory was first proposed in 1838 by Theodor Schwann, and is usually attributed to him and Mathias Schleiden. Though it took the rest of the nineteenth century to clarify its implications, by the century's end, the common nature of all life-forms—animals, plants, protozoa, bacteria—how life was reproduced, and where it comes from was understood with relative clarity. It is not overstating things to say that without cell theory, modern biology would not exist.

Looking back, we can see that with these three magnificent possessions in hand, the path to the modern understanding of life was secured. Though there were many barriers to overcome— some immense, others unimagined—with Newtonian physics, modern chemistry, and cell theory, and with the requisite intelligence, courage, and perseverance, humankind's quest for a deep understanding of the nature of life would be fulfilled. However far in the distance, there was light at the end of the tunnel. And at the beginning of the nineteenth century, the call for an independent discipline of biology, separate from physics and chemistry, was realized.

But paradoxically even as it separated, biology drew closer to physics and chemistry, and a new experimental biology, physiology, was born whose mission was to understand life by applying the methods and ideas of *physics and chemistry* to living things. Still, to freely travel the path ahead, a critical intellectual battle had to be waged, and an old dialectic confronted hopefully for the last time. The debate that ensued acquired great polemical intensity, not merely because of conflicting scientific ideas, but also because there was a lack of precision about what was actually being disputed, and perhaps most important, because what was at issue seemed to have significant religious implications.

With much historical precedent, one side claimed a special force, substance, material, essence, soul, spirit, property, or principle for life that followed rules distinct from those that govern inanimate objects. This claim subsumed two related propositions: First, living things contain certain elements (as Aristotle referred to them), forces and substances that are unique to them. Second, these elements obey rules or laws that are different from those for inanimate objects.

The other side, bolstered by nineteenth-century physics and chemistry, maintained that both of these propositions were false; that no special substances, forces, or laws existed for living things. They claimed the antithesis. Living things and their properties are manifestations of the same substances, forces, and laws that operate in the world of inanimate objects. And so, one side argued for a unique physics and chemistry of life, while the other denied its existence, and believed that all material objects, whatever their nature—animate or inanimate—are comprised of the same sorts of substances, respond to the same forces, and obey the same laws of physics and chemistry.

Though the less famous of the two great nineteenth-century debates about life—the more celebrated being the debate about evolution—the debate over *vitalism,* as it came to be known, was no less important. Those who believed in special forces, substances, and laws were called *vitalists*, and those who did not were called *determinists, materialists,* or *physicalists.*

As suggested, this debate was not new. It had begun in ancient Greece, had been pursued vigorously by Roman Catholic clerics for a thousand years, and was an important aspect of the intellectual explosions of the Renaissance and the Enlightenment. In the nineteenth century, at the dawn of the modern era, it once again took center stage. But there was an important difference: this time there seemed to be a good chance that the dispute could be resolved.

The disagreement had important religious overtones and imitated similar arguments about life's nature that had taken place over the centuries. The most famous of which was Descartes' rejection of life's sublime essence (with the exception of the human soul) and his radical assertion of its mechanical nature. However cautiously and reluctantly, he rejected the religious tradition of Christianity, that viewed life, especially human life, as something apart from the rest of the natural world, created in all its aspects and manifestations, in all its wonder and variety by God in ways beyond our understanding.

Since vitalism proposed special life properties, to some it appeared to be offering a religious explanation for life, while others thought that the materialists were imagining a godless world bereft of meaning. This argument was much the same as the roughly contemporaneous debate about evolution. It was thought that science and religion were incompatible and irreconcilable opposites, and one had to choose between them. For this reason, the debates were not merely dialectic, and often generated far more heat than light. Even today, the vitalist dispute is often mischaracterized in black-and-white terms as having been between those who held magical beliefs in supernatural forces (and as such had to not just be shown wrong but discredited) and those who cleaved to the rigorous science of materialism.

In any event, the debate lasted for many years and had a natural ebb and flow. From time to time, new evidence shifted the balance in one direction or the other. It drew into its cauldron—sometimes unwittingly, sometimes unwillingly—the most prominent biologists of the time. Among them were devoutly religious vitalists, atheistic materialists, devoutly religious materialists, atheistic vitalists, as well as agnostics on both matters.

Though not the common impression then or today, most vitalists did not believe in immaterial forces, just unknown ones. While on the other hand, some in the materialist camp held powerful vitalist notions of their own because so much about life seemed to defy simple physical and chemical explanation. It was not uncommon for a scientist to seem to be on both sides of the debate at the same time, to appear to change sides with each utterance, or to think he was on one side, while his colleagues thought he was on the other. And one believer's vitalism or materialism was not likely to be the same as that of the next. The often-acrimonious dialogue made sorting things out all the more difficult. But as the years went by, to call someone a *vitalist* became a curse. It implied that you were ignorant, not to mention imbecilic, reactionary, and even evil.

From today's perspective, it seems that the materialists had nature on their side, and the vitalists' world was illusory and, to use a choice word, insubstantial. However, the outcome of the debate was far from obvious at the time. One side did not have all of the ammunition. Yes, the materialists had physics and chemistry on their side, but there was so much about life that even in their most fanciful imaginings they could not explain.

## Pasteur

Let me illustrate the conflict with two seminal experiments by the great nineteenth-century French biologist Louis Pasteur that led him to opposing conclusions about life's nature—one materialist, the other vitalist. The experiment that led to a materialist conclusion concerned the polarization of light by certain organic molecules. The chemist Mitscherlich had shown that two chemically identical substances—tartaric and racemic acids—found in wine fermentation vats displayed different optical properties in solution. Tartaric acid twisted polarized light in a clockwise direction, while racemic acid was without effect.

Pasteur thought this strange. Why was the essentially identical racemic acid without effect? Though only twenty-two years old at the time, his great perspicacity was already evident. Looking at crystals of racemic acid in a microscope, he noticed that there were two kinds, one the mirror image of the other. He set out to sort them by hand, crystal by crystal. Having separated them, he then tested their optical properties and found that molecules from one crystal rotated polarized light in a clockwise direction, whereas those from the other rotated it counterclockwise. When they were together, each quenched the effect of the other. He called this phenomenon *disymmetrie*.

What for another person would have been a straightforward chemical observation, to Pasteur suggested nothing less than a chemical explanation for life. He thought that two seemingly identical substances displaying these basic differences provided a deep insight into the nature of life. He knew that such optically active substances were unique to the living things that produced them (yeast in the fermentation vats). Consequently, he believed, and continued to believe throughout his lifetime, that these disymmetric forces were a reflection of life's underlying cause. And since dissymmetry was a property of chemicals, life had a chemical or more generally a material basis.

The experiment that led to a vitalist conclusion is one of biology's most celebrated. It concerned the spontaneous generation of life from inanimate matter. Pasteur put the final nail in the coffin of this idea held for centuries by many scholars. With a prize from the French Academy of Sciences in the offing for a definitive experiment, Pasteur sought to resolve a dispute between two Catholic clerics—Joseph Needham in Britain and Abbe Spallanzani in Italy—that had taken place a century earlier.

The story began even earlier. One hundred years before Needham and Spallanzani, an Italian poet, physician, and naturalist named Francesco Redi performed the following experiment. He placed fine gauze between the air and a decaying piece of meat. His intention was to determine whether the maggots that formed in and on the meat came from the meat or the air outside. With the gauze in place, no maggots were seen. He concluded that the lifeless meat had not generated the maggots spontaneously, but that their cause was external, from the air (as we know, from flies).

Still some thought that if not maggots, more primitive life might be spontaneously generated from matter. This was the possibility that Needham and Spallanzani tackled. Needham concluded that indeed some sort of primitive life was generated spontaneously. Even after heating a meat broth in sealed containers, living objects (bacteria) appeared and clouded the water. Spallanzani was unable to repeat Needham's study. When he heated the fluid for a longer period of time than Needham, he found no clouding, no signs of life.

Needham and Spallanzani agreed that the conflicting results were caused by differences in the duration of heating. Needham said that Spallanzani's experiment failed because of excessive heating, while Spallanzani said that Needham's was the result of insufficient heating. Into the breach, a mere one hundred years later, came Pasteur. He clarified the situation, and did so very simply. He added something that was not all that common at the time—a control experiment. While he protected one previously heated flask from external, airborne contamination, another was intentionally exposed—the control. The broth in the former vessel remained lifeless, as Spallanzani had found, whereas life appeared in the latter as reported by Needham.

Pasteur concluded from this that Spallanzani was correct, that Needham had not heated his samples long enough to kill the bacteria. Life, he deduced, came from an external airborne source, not the meat in the broth. In its time and context, and as understood by Pasteur, this was a vitalist conclusion. Though we know today that life was introduced by contamination with airborne bacteria, when Pasteur performed the experiment, all he could say with certainty was that a vital property had entered the flask from the air and produced life. He saw the experiment as a test between vitalist and materialist views, a test that vitalism won.

## A Profound but Incomplete Victory
## for Materialism

But we can easily turn the conclusions of both of Pasteur's experiments, one an apparent victory for materialism and the other for vitalism, on their respective heads. If dissymmetry is a unique, even causative feature of life, then might we not argue that it indicates the presence of special chemical laws that apply only to living things, a vitalist conclusion?

As for spontaneous generation, we now know that the growth of new life in the contaminated flask was the end product of a complex chemical process that began with the duplication of DNA in bacteria, and that this is a chemical or materialist event. We can even turn this conclusion around one more time if we wish. We can say that the chemical DNA contains a genetic code that follows rules unique to living things found nowhere else in the material world, and isn't this a vitalist conclusion?

In a modern light, the dialectic between vitalism and materialism collapses. We now know that distinctions that seemed so powerful at the time were at least partly illusory. We have learned that there are chemicals that are unique to life, that follow rules that are unique to life, and yet there is no special life chemistry, just chemistry. We have learned that the chemicals of life and their reactions are *vital* examples of the same laws of chemistry and physics found in the world of inanimate objects. Vitalism and materialism, it turns out, were not dialectical opposites, but surprisingly were conceptually compatible. Life's special chemicals were evidence of both vitalism and materialism.

Nonetheless, the debate had a winner and loser, and the victory was as clear as could be. The materialists won the battle convincingly. So much so that vitalism became a permanent term of opprobrium. The lasting judgment was that whatever its source and nature, life is based on the same laws and principles, the same chemistry and physics, found in the world of inanimate objects.

This was an extremely important conclusion, but it was negative. It told us what life is *not*. There is no special life physics or life chemistry. But no *affirmative* understanding of life emerged. Despite establishing its physical and chemical basis, no explanation of what life *is* dawned. Was it attributable to its materiality, a question of organization, or was there something else as yet undiscovered? After all was said and done, though the vitalist debate told us what life is not, it left the central question of biology—beyond cells, what is it that makes something living?—unanswered.

CHAPTER 2

# The Material Causes of Life

## *The Modern Perspective*

———

[The living organism] is the finest masterpiece ever achieved
along the lines of the Lord's quantum mechanics.
**—ERWIN SCHRÖDINGER, *WHAT IS LIFE?*, 1943**

MUCH AS the discovery of the atom led to a search for sub-atomic particles, after the advent of cell theory, one reduction led to another. If the cell was alive, then it was natural to ask what material within it was living? What substance animated it?

Two hypotheses held sway during the latter half of the nineteenth century. The first viewed aliveness as a physical phenomenon, like water running or chemicals crystallizing. Attention focused on protoplasm, the seemingly uniform substance that filled the cell. Protoplasm oozed, flowed, streamed, contracted and expanded, and seemed very much alive. When the cell's enclosing membrane was punctured with a tiny needle, a slimy material could be seen escaping from its constraints as if alive. For those who sought to comprehend life's nature from a physical perspective, coming to understand the lifegiving properties of this protoplasmic ooze, either as a whole or as an agglomeration of yet smaller basic particles (for example, Darwin's Gemmules) was paramount, because it was life's ultimate source.

The second hypothesis followed Lavoisier. It looked for life in the chemical substances of cells and their reactions. For those with this viewpoint, extracting the chemical components of the cell, separating them from each other, and identifying their structure and chemical properties was the essential task. Unlike scientists seeking a physical explanation for life, they were not looking for a uniform living material, but to the contrary, for variety—for different chemical substances that did different things. They did not seek to generalize, but to specify and categorize.

This approach required the invention of all sorts of chemical and physical methods to extract, separate, and identify molecules. The techniques were usually empirical, and often made use of the ways of the cook. By grinding up and extracting solid animal tissues, such as muscle and liver, or collecting blood, it was possible to identify a variety of substances. These workers were often referred to disdainfully as "kitchen chemists" or "grind and finders." But however disparaged, they labored away, finding all sorts of chemicals, acids and bases, sugars and amino acids, proteins and enzymes, and on and on. For them, the key to understanding life's nature was learning about the reactions and interactions of these molecules.

Those who sought a physical explanation for life called themselves *general physiologists* (today's biophysicists and cellular physiologists), while those who looked to chemistry for answers called themselves *physiological chemists* (today's biological chemists or biochemists). In the end, the research program of the general physiologists failed. Though they characterized the properties of protoplasm using contemporary physical principles, nothing about those properties seemed to explain life. Indeed, the physical properties of the protoplasm seemed much like the physical properties of various inanimate fluids and gels, various liquid and solid states of matter, and were often compared to them. In a sense, this was exactly what was expected. After all, if there was to be a physical explanation for life, didn't it have to be based on the same physical properties that were found in the inanimate world? But this left the general physiologists unable to answer their key question—what about this protoplasmic material made cells alive?

By relatively early in the twentieth century, it had become clear that life could not be attributed to any single living substance or particle. Eventually and inescapably, it had to be accepted that there was no such thing as protoplasm. The cell was not filled with a single lifegiving substance, but contained a complex pot-pourri of chemicals involved in all sorts of reactions, most importantly, or so it seemed, those concerned with energy utilization and conservation.

Even though the chemists were no more able to explain life than the general physiologists, they were correct about what cells contained. They contained chemicals, not protoplasm. And consequently it became unmistakable that to understand life's nature from a materialistic point of view was a problem, an incredibly difficult and complicated problem, in *chemistry*, not physics. The cell was a mass of chemical substances that carried out life's activities and that as such, were lifegiving. The result of this understanding was that identifying the chemical basis of life became a major task, if not the single major task of twentieth-century biology.

The hope was that if they could establish the nature of life's chemistry, they would come to understand its lifegiving attributes. Indeed, as said today many scientists believe that we have done just that. Though we cannot yet catalogue all of life's chemicals, or detail their structures, reactions, and interactions, the central chemical features of life are well known, and can be described with some precision and breadth. Most remarkably, a large and influential group of scientists have come to the conclusion that life can be understood, or at least understood in all significant respects, in terms of the properties of just two chemical substances: DNA and proteins.

## From Genotype to Phenotype

The journey of genetics—the study of critical but initially obscure things called "genes"—that began with Gregor Mendel's famous pea plant in the mid-1800s came to a successful conclusion in the 1950s with the discovery of the location and character of the gene in the famous double helix of the DNA molecule. Geneticists had already shown that the gene, the genotype, was the underlying cause of life's observable realities, its appearance, features, and functions, its phenotype. It seemed that all of our characteristics—from eye color, to height, to how we digest food—was based on genes. And genes turned out to be surprisingly simple things, nothing more than chemical structures, sequences of molecular subunits within the large DNA polymer. The implications of this discovery were truly astounding. If the gene was in DNA, immanent to a chemical substance, then it seemed so was life itself. Life was incarnate to a molecule!

But the gene was not a substructure of life, like an atom in a molecule. Rather, it was a chemical code that contained information for the construction of another type of molecule—proteins. Proteins came in various forms, and genes determined that form. It was through the actions of protein molecules, not genes, that life's various aspects were actually realized. DNA provided the information and proteins gave rise to life's observable features. And so, it seemed that three simple statements could explain the phenomenon of life:

- ▸ The information needed for the expression of all of life's varied characteristics is found in the most extraordinary DNA molecule.
- ▸ This information is for the structure of just one other type of molecule—proteins.
- ▸ The multiplicity in form and function of protein molecules gives matter life.

## Chemical Immanence

From this perspective, life is immanent, inherent, and attributable to a particular chemical or group of chemicals. There are two variations on this theme: In one, special molecules, DNA or alternatively DNA and protein (not Pasteur's polarizing acids), are endowed with the necessary and sufficient properties to produce life. In the other, the rest of life's chemicals are added. When taken together, they achieve the same lifegiving end. These chemicals are the various substances that are essential for cellular and organismal life—substrates, products, and catalysts. They are concerned with energy metabolism; organic synthesis; degradation; oxygen utilization and carbon dioxide production (vice versa for plants); movement and responsiveness, including the movement of various substances into and out of cells for nutritive and osmotic purposes; and last but certainly not least, reproduction, growth, and cell division. In this view, it is the chemicals involved in these events with their multitudinous reactions and interactions that vest particular objects with life.

And yet when all was said and done, how chemistry yielded a living thing was not clear. However one looked at it, biological chemistry, even DNA chemistry, looked suspiciously like, well, chemistry—certainly a complex and unique chemistry, but nonetheless like chemistry—not life. To many this was the point. We had learned that life is chemistry.

But to others, this simply begged the question: how does chemistry yield life? How does life arise from its chemistry? To skeptics, if we could not answer this question, then how could we say that chemistry explains life? How could we say that chemistry explains life, if we could not specify in what way it produces it?

CHAPTER 3

# Coming into Being

## *The Origin of Aliveness*

———————————— ∽ ————————————

Probably all the organic beings that have ever lived on this earth
have descended from some one primordial form, into which
life was first breathed.

**—CHARLES DARWIN, *ON THE ORIGIN OF SPECIES*, 1859**

ONE WAY to approach the question of what makes a particular
material object alive is by asking about life's origin. Can we identify the transformation that made some primordial object alive?
What property, what change of state first bequeathed aliveness?

When we refer to life's origin, we usually think of a particular
event or series of events that took place some four billion years
ago that gave rise to the first living things on this planet. With
Darwin, this is traditionally imagined as a singular occurrence,
unique and discontinuous, that divided the world that existed
from the one it brought into being.

Before life's origin, the earth could be fully described in terms
of its geology—its rocks and rivers; its solids, liquids, and gases;
its heights and depths. Though roiling with activity and reactivity,
all was inanimate. Afterward, one could not describe, much less
understand, earth without reference to its living forms, its animate beings. Though we have no grounded idea of what chemicals were involved, or what the actual events of life's origin were,
at some point in time a particular object, or more likely a group

of like objects, came into being that were the elementary living thing, the first species.

If first life was as Darwin imagined it—singular, monochromatic, and like Adam, solitary—the evolution that followed was kaleidoscopic. Evolution was an elaboration on the simple ancestral objects and provided multitudinous variations on their theme. In this regard, scientists have traditionally envisioned the history of life in two distinct phases. First, it originated, that is, came into being, and then it evolved, or underwent changes in form and substance.

The embellishment and expansion that evolution wrought was extraordinary and eventually left the earth's surface covered with a remarkable abundance, and an even more remarkable variety of life forms, along with their detritus. But for all of life's incredible variety, for all its forms and variations, according to this view, evolution created nothing *fundamentally* new. It was nothing more than an enormous elaboration on the theme of aliveness that had been established with first life. Evolution reworked, extended, and expanded upon the original living thing in many varied and wondrous ways, but the property of "aliveness" that had come into being when life first appeared remained unchanged. Whatever it was, it was present in all living things regardless of their particular incarnation.

In contemporary times, the idea that life came into being as the result of some particular singular event is often dismissed. In this view, life came into being gradually, not all at once. Some primitive chemical systems were more evolved than others and as such could be said to be more alive, more animate. But no sharp dividing line could be drawn. There was no clear demarcation between the inanimate and the living, only a continuum, as life slowly amassed and aggregated.

But, viewing life's origin as gradual does not give us license to sidestep the need to differentiate the animate from the inanimate. The reason is self-evident. If we say that the living and the inanimate are different, we must explain the difference. And this difference must be due to some particular specifiable feature or group of features that makes one object alive, while leaving another, lacking it or them, inanimate.

Without the ability to make this distinction explicit, the phenomenon we call life becomes obscure and unintelligible. Though we would still be able to observe and catalogue its various features, and could intuitively understand that the difference between the animate and inanimate is real in every imaginable respect, our understanding would not be amenable to the ablutions and ministrations of science. However difficult, however inconvenient, if we hope to study life's origin as a matter of science, we must be able to articulate what event, what property, what state, what substance first separated the inanimate from the living. Some sort of evolutionary gradualism cannot cloud this requirement.

## Reproduction as Life

And so, whatever our desire, we must ask what event first introduced life to this planet (or elsewhere in the galaxy)? If there is a common answer to this question among scientists today, it is *reproduction*, the ability of an object to replicate itself. In this view, reproduction was and remains the lifegiving property. When it appeared, so did life. Moreover, reproduction sustained life, and allowed the extremely fragile objects that living things are to survive for four billion years and counting in spite of the enormous forces of decay and destruction that constantly bombard them. This is because reproduction creates new life, just as ceaselessly as the forces of dissipation end it.

Nor could life have evolved in its absence. It is an essential substrate for evolution. Because of reproduction, as well as genetic mutations, nature has various products to select among, determining which survive and which do not and thereby setting the course of evolution. As a consequence, we can say that reproduction is in no small measure responsible for the immense diversity we see in the form and function of living things. Life's variety has been in great part created through its agency. While our off-spring testify to our success as parents in propagating our own kind, at the same time our children serve as instruments in the continuing struggle for survival—testimony to the centrality of reproduction to life.

Finally, and critically, reproduction is unique to living things. To our knowledge, comparable events do not happen in the world of natural inanimate objects. Given these imposing facts, it is easy to see why one might conclude that reproduction is nothing less than life's embodiment—what makes some particular material object alive.

In this case, from a materialistic and reductionist perspective, we are obliged to explain reproduction in basic physical and chemical terms. That is, despite its uniqueness to living things, we must be able to reduce reproduction to general, nonbiological phenomena that obey the same laws of physics and chemistry as everything else. Though reproduction has many complex forms and guises—from cell division to human pregnancy to the transfer of pollen in plants—amazingly it can be understood in *chemical* terms that are common to them all. Indeed, it can be understood in terms of a single chemical property of a single molecule—the replication or duplication of DNA, the manufacture of new, identical DNA molecules from its precursors.

Once again, we find ourselves looking at DNA. It turns out that DNA creates life in two ways, first in the expression of its subsumed genes as proteins, and second, in its duplication. We can say that the latter makes new life, while the former makes life real. Either way, as protein expression or DNA duplication, as production or reproduction, as maintenance or origination, it is the properties of DNA that account for life.

And so, while there are many differences between species in the particularities and peculiarities of how reproduction occurs and how a new organism is produced, at the most fundamental level, it is the copying of DNA (and RNA in some viruses) and its transmission from parent to progeny that creates, propagates, and amplifies life. As already explained, however demeaning to the grandeur of our species, it seems that everything about life, or at least everything pivotal, can be understood at the deepest and most profound level in terms of the properties of this single, singular molecule. Life both arises and is maintained through its agency.

## *Life beyond Reproduction*

Putting aside DNA as life's residuum, however critical, however essential, reproduction cannot be equated with life and its origin. This fact is obvious, even self-evident. But before I explain why, let me let you in on a little secret. Scientists have struggled unsuccessfully to find a molecule in living things that is actually capable of reproducing itself, including DNA. Though such autocatalytic substances have been produced in the laboratory, to the best of our knowledge no such naturally occurring chemical entity exists on the planet today. Nothing happens in a test tube filled with DNA and the entourage of precursor molecules needed for its duplication. They just sit there, inert.

Special protein catalysts called DNA polymerases are required for its duplication. This presents a conundrum, a chicken and egg problem. DNA can only be replicated in the presence of DNA polymerases and DNA polymerases are proteins, the product of a DNA code, and therefore can only be produced in the presence of DNA. It seems that DNA replication could only have occurred in the first instance if DNA polymerases already existed, and DNA polymerases could not have existed without preexisting DNA to provide the needed code. There have been several attempts to deal with this problem, but they have not really dealt with the paradox, or eliminate the contradiction. Simply and unavoidably, this can only be done if protein manufacture (the synthesis of DNA polymerases) *predated* reproduction (DNA duplication). In more general terms, unless it occurred without catalysis, whatever chemicals were involved in the original form of molecular duplication required that the catalyst be present first.

In any event, setting aside such confusions, in spite of the centrality of reproduction to life and its likely presence in the penumbrae of its origin, it cannot be equated with life. The reason is plain enough. At any given time, most living things are *not* reproducing. The human female after menopause and the male after a vasectomy, though unable to produce offspring, certainly seem alive. Indeed, the activities of nonreproducing organisms, as well as nondividing cells (cells that are not replicating their DNA), account for most of those we associate with life. Because organisms are alive when they are not reproducing and even when they are incapable of reproducing, the equation of life with reproduction fails. However central, and it is central—life cannot evolve or be sustained in its absence—reproduction is not necessary for life's attendance, and as such cannot explain its creation.

In an essay in *Science* magazine a number of years ago ("The Seven Pillars of Life," 295 [2002]: 2215–16), the late Daniel Koshland, a former editor-in-chief of the magazine and at the time a professor emeritus of biochemistry at the University of California at Berkeley, described an earlier "conference of the scientific elite" in which a "definition of life" was sought. As Koshland reported it, after much discussion the group seemed to make a determination. One "statesman" spoke up and expressed the consensus. "The ability to reproduce," he said, "that is the essential characteristic of life." But this agreement was short-lived as "a small voice" said, "Then one rabbit is dead. Two rabbits—a male and female—are alive but either one alone is dead." This curious characterization, somewhat reminiscent of Schrödinger's cat, seemed to change most opinions in the room and the assembled scientists apparently realized that life could *not* be defined by reproduction.

But having rejected this explanation, they came to a disconsolate and pitiable conclusion. It was not a new conclusion; indeed, as mentioned, it had a storied history. It seemed inescapable to them that however it began life just could not be defined. They resolved, "There is no simple definition of life." Yet, they continued, "everyone knows what life is." But what is it that everyone knows, and yet cannot define?

CHAPTER 4

# A More Refined Understanding

## A Modern View of What Is Necessary for Life

———— ⌒ ————

Life is combustion.

—ANTOINE LAVOISIER, *OEUVRES DE LAVOISIER*, VOL. 2, 1862

AND SO, reproduction and its underlying basis—molecular duplication—are not the keystones of life because life exists in their absence. But in this case, what first separated the world of the living from that of the inanimate? What change introduced life? Perhaps, as has been variously suggested, there was a chemical synthesis that did not involve molecular duplication, say a reaction that produced energy, or used it, or maybe life's coming into being was due to a physical event such as the performance of work, or perhaps it was created by the separation of certain molecules from their environment by self-aggregation or by a barrier to diffusion like a membrane.

But there is a serious problem with looking for the source of life's creation in such physical and chemical properties. The difficulty is that if we do, we are then obliged to ask what about these properties is life entailing. That is, we must explain how and why they produce life. As it turns out, despite our great understanding of the physico-chemical nature of living things, such an explanation is not possible.

In light of the fact that it has been shown over and over again to the point of obviousness that living things obey the same laws of physics and chemistry as the rest of the material world, and that there is no special life physics or life chemistry—just physics and chemistry—to lay life at the door of such processes, we must explain what about them transcends their inanimate antecedents; what about them introduces life? Even the *informational* content of DNA's genetic code, though certainly a unique feature of living things, affords no opening. Indeed, it provides a good example of the general problem. To view the genetic code as lifegiving, we have to explain why a particular sequence of chemical substances (the nucleotides that comprise the DNA polymer) that provides a code for the structure of another type of molecule—proteins— makes something alive? How does it differ from any other code that we might imagine encrypted in some *inanimate* substance, except in its particularities?

It is sometimes thought that this difficulty of explanation and extrapolation can be overcome if we view the lifegiving feature as composite. Life is not due to a single chemical or physical event or substance, but to a combination of things. As Koshland's scientists said, there is no simple definition of life, but there is a complex one. Yet how does this change the situation? In what way does putting a bunch of chemical reactions and physical events together help matters? Why would a *group* of them be any more lifegiving than one alone? What about their coming together, about their community, causes them to rise above their lifeless precursors to make them alive? It must be stated forthrightly that if there is anything to this contention it cannot merely be claimed, it must be proven. It must be shown on what basis these phenomena as seen in living things differ in kind from combinations of chemical reactions and physical states otherwise.

Some scientists who believe seeking such proof is futile have looked to life's organization, its information content, its complexity, that is, to Aristotle's life-as-organization, for its essence. But if we are unable to explain how life arises from chemistry and physics, we seem no more able to explain how it arises from its organization. Why does the organization of something make it alive? What about it is lifegiving? How does the organization of living things differ, and differ *qualitatively*, from complex organized systems that exist or that we can imagine that are not alive?

Like the opinion that life arises from the common presence of certain physical and chemical features, it is sometimes suggested that it is a matter of the *degree* of complexity. But how would this work? While it is certainly true that life is a uniquely complex phenomenon, why would any particular degree of complexity in and of itself make something alive? Can't we at least imagine an entity as complex and well organized as a living thing, an automaton of our creation, that is wholly inanimate? What would the difference be? And what degree of complexity or organization produces life? Is there a borderline? What is too little, and is there such a thing as too much? If we hope to prove that life can be understood as an extreme case of complexity, then, as with physico-chemical explanations, we must provide convincing answers to such questions.

In any event, whether as physics, chemistry, or organization, or all of them together, thus far science has been unable to provide an affirmative explanation for how life arises from its inanimate antecedents. All we have are unsupported claims and unproven suppositions—something unknown and seemingly undecipherable occurred that introduced life into a purely physical world. The unavoidable truth is that however we twist the notion of aliveness, and despite the fact that science can provide clear answers to what life is *not* and has purged insubstantial explanations such as life forces and vital substances from its lexicon, in the end it has been unable to provide an affirmative statement of what precisely life *is*. If life-as-chemistry and physics is just chemistry and physics, and life-as-organization just organization, and if both taken together are no more than each alone, then don't we with all of our knowledge find ourselves in much the same situation as the benighted nineteenth-century scientists embroiled in the vitalist controversy, with the phenomenon we call life beyond our ability to comprehend?

## The Modern Materialist's View

To the modern materialist, all this hand-wringing is pointless and naïve. We are looking for a *fundamental* distinction, a lifegiving feature, where one does not exist. We do not have to discover some mysterious quality emerging from life's physical and chemical incarnation to separate the living from the inanimate. It does not exist. There is no essential difference between the chemical and physical embodiment of living things and that of inanimate objects.

The first living thing did not differ from its immediate inanimate precursor in any fundamental way. It was unlike it and its neighbors all right, but only in its *particularities*, in its particular incarnation. But these were ordinary physical and chemical differences, no different in kind than those between various *nonliving* chemical and physical systems. Life from this point of view is to be found in its distinctive, though quite ordinary, physical and chemical features, not in some undiscovered, indeed unspecifiable, emergent state of aliveness.

It is to be found in the material particularities of living things. They and they alone give life. There is nothing else, and nothing else need be imagined. Simply, when certain chemical and physical structures and states become manifest with certain organizations of matter, life comes into being. Together they define life. They make it real. Whether we are talking of first life or of life thereafter, its essence is to be found in certain distinguishing chemical substances, reactions and interactions, in specified physical states, and in a particular organization. According to this view, life and its material embodiment are, as said, two sides of the same coin, one in the same thing. This was so at life's inception and remains so today.

Given this understanding, the task of science is to learn as much as possible about life's material *particularities* by deciphering the complete structure of DNA in humans and other species, for example. When these distinguishing material features of living things are known in all necessary detail, life's essence will have been fathomed—its essence and its particularities being one in the same thing. The great appeal of this point of view is that it appears to relieve us of the inconvenience—though it is more like an immense and overarching difficulty—of having to account for and describe some unknown and seemingly unfathomable emergent state of aliveness. Such a state does not exist, and need not be sought. Life's material particularities and life itself are identical—that is—the same thing. This is not only all that we need to know, there is no more to be known.

But however convenient and simplifying, this is just not so. As said, claiming an identity does not get us off the hook. It does not relieve us of the need to provide an explanation for life. All it changes is the nature of the explanation that is required. While it is true that if an identity between life and its material particularities can be assumed, we do not have to explain what lifegiving feature or features emerges from its material causes, we must still specify what chemical reactions and what physical and organizational states produce life. If first life and life since can be defined by the presence of certain material features, what are those features? If we hope to affirm this point of view, we must be able to supply such a list and furnish such an accounting. Otherwise, all we have is the *assumption* that life is identical to its material embodiment, and however convinced we may be that this is so, it is not science.

Can this be done? As explained, science has little of substance to say about first life, save a variety of often fascinating speculations, having no record of its origin, much less knowledge of the particular chemistry and physics involved. But we have a great deal of painstakingly accumulated knowledge of the chemical, physical, and organizational properties of life as we find it today. And if life can be defined by its material particularities, if its particularities and its essence are indistinguishable, then we would seem to be in a very good position to assemble the essential list. We can call it the list of life's *general* particularities—those physical and chemical circumstances and states that produce life that are common to all living things, but are not found in the world of inanimate objects. That is, what is it that everyone knows, but cannot define?

## PICERAS

In his essay "The Seven Pillars of Life," Daniel Koshland provides his personal list of these features in the form of a Greek-sounding acronym—PICERAS. PICERAS seems to take us beyond the simple equation of life with DNA and the genetic code, and even beyond life as chemistry. As such, it provides a more refined understanding of the phenomenon. Let's go through it, one letter at a time:

- ▶ "P" stands for the "program," for an "organized plan that describes both the ingredients (of living things) themselves and the kinetics (timing) of (their) interactions." The program is found in DNA—"in DNA . . . the program is summarized and maintained." The program is the genetic code and the means used to regulate its use.
- ▶ "I" is for "improvisation." Improvisation refers to the agency or agencies that change the program. There are two: random changes (mutations) in the structure of DNA and the action of natural selection on the particular DNA program. As such, improvisation is an evolutionary property.
- ▶ "C" is compartmentalization. Whether we consider cells separated from their environment by their membranes or organisms by their skin, living things, though not totally separate and distinct from what surrounds them, are discrete objects; that is, they are compartmentalized.
- ▶ The fourth pillar is "E" for energy. Koshland relates the need for energy to movement. Certain chemical reactions and physical events make energy available for the *directed* movement of molecules and objects, as opposed to their random or undirected movement.
- ▶ Fifth, is "R" for regeneration, which Koshland describes as the need to replace lost substances and components. In the normal course of events, the constituents of life are lost and

need to be replaced, and the mechanisms responsible for their replacement are those of regeneration.

▸ Pillar number six is "A" for adaptability. Koshland is not referring to Darwinian adaptations here, but to the ability of organisms to respond to changed circumstances by means of *feedback* mechanisms, such as the "behavioral response to pain" or the creation of conditions such as hunger and satiety. These feedback mechanisms are also Darwinian adaptations, but that is a subject for a later chapter. Briefly though, through them living things are able to sustain certain stable states despite the challenge of environmental variation. For example, using feedback mechanisms, mammals (homeotherms) are able to maintain a constant body temperature when faced with a changed climate. As we shall see, feedback is critical for producing "homeostasis," the constancy of the organism's internal state.

▸ The last pillar is "S" for seclusion. For living things to be sustained, their various reactions and interactions must be separated from *each other*. Seclusion may either be the consequence of the specificity of the reactions or their physical isolation. This may seem odd, since without contact among life's substances, there can be no life. Nonetheless, contact must be controlled and specific. For that, chemical or physical seclusion is required. Otherwise, all order would be lost and molecules could interact with each other willy-nilly or not at all.

## *Necessary and Sufficient*

Though the PICERAS acronym takes us beyond a simple chemical and physical materialism, nonetheless the list remains deeply materialistic:

▸ The program is in the DNA molecule and its chemical expression.

▸ Improvisation is a change in molecular structure and the result of that change.

- Compartmentalization involves the physical separation of chemicals in space.
- Energy to do work is obtained by chemistry.
- Regeneration refers to chemical events and their physical sequelae.
- Adaptability refers to physical and chemical feedback mechanisms.
- Seclusion involves the physical or chemical isolation of molecules and reactions.

Though he makes a passing stab at considering the life of complex organisms, the list is really what Koshland thinks of as the lifegiving properties of *cells*. It has nothing to say about life as complex organisms live it beyond very general notions like compartmentalization and regeneration. But let us put this concern aside for the moment, and for the sake of discussion say that each item in PICERAS, or some similar list that we might compile, is necessary for life in all its forms and that there are no other prerequisite material features. With this list in hand would we have answered the ultimate question—"what is it that makes something living?" PICERAS!! After all, according to the materialist claim, if all of its enumerated features were present, we would expect to find life. If it were the true and essential list of the material properties necessary—essential and indispensable—for life, the presence of its members would make something alive.

From this point of view, if we can provide such a list, if we can enumerate all of life's necessary features, then we will have explained the phenomenon of life, and our job will be done. But if you think about it for a moment, this is a dreadful mistake. Such a list, however well wrought, only tells half the story. Yes, it provides knowledge of the features that are *necessary* for life, but it tells us nothing about its *sufficient* properties, those properties in whose presence the thing in question is guaranteed. Regarding life, they are properties or features in whose company life invariably exists, whose presence tells us that something is alive. This is as opposed to its necessary properties whose presence, though required for the thing in question to be realized, guarantees nothing. As such, in the end, all we have in such a list is what is necessary for life, with no idea of what guarantees its presence.

But, the materialist may protest. Though it may seem otherwise, this list provides us with both life's necessary *and* sufficient properties. The two are inextricably joined, just as life's physico-chemical embodiment is joined to life itself. They are one in the same thing. The list is a list of both.

Nevertheless, this conclusion cannot bear scrutiny. A DNA program, though certainly necessary for life, does not guarantee its presence. DNA and its program can be found inanimate in test tubes in research laboratories all over the world, as well as in dead bodies. And things can certainly be compartmentalized, require energy and be secluded, and not be alive, even if these features are required for life. Even the very biological sounding regeneration is as suggested at base a simple chemical principle. Reactions generate, and they also regenerate, and do not have to be alive to do so. If we think of regeneration as replacing lost limbs or healing wounds due to cell division, that is, as the result of (cellular) reproduction, then, as we have discussed, objects can be alive in its absence.

This leaves improvisation and adaptability. The word "improvisation" seems to connote something like free will, an independent choice internal to the object. But as explained, this is not what Koshland has in mind. He points to two kinds of improvisation. The first is mutations in DNA. Mutations are caused by random physical and chemical events, such as cosmic radiation, that change the structure of the DNA molecule. Though necessary for life's evolution, all mutations guarantee is the presence of an altered molecule, not life.

The second source of improvisation is natural selection, the effect of external environmental forces on the fate of things—in Koshland's formulation, the fate of the altered DNA molecules. But *all* material bodies, not just living ones, are changed differentially, in form and substance, by such forces. That is to say, natural selection is a feature of the whole material world, not just living things. And so, though along with mutations it is the driving force of the evolution of life, also like it, it does not guarantee life's presence in and of itself. Indeed, both mutations and natural selection are examples of *nature's*, not *life's* improvisation.

Finally, there is adaptability. As said, Koshland thinks of adaptability in terms of feedback mechanisms, like those exhibited by the thermostat on my wall. Putting their role as Darwinian adaptations aside, not only thermostats, but all sorts of inanimate machines—from television sets to clothes washers to automobiles—exhibit feedback. And so, though feedback mechanisms are a critical aspect of living things, their presence does not guarantee life's presence. They are not a sufficient cause of life.

## Life as Chemistry

As it turns out, none of the elements of PICERAS—not genetic programs, improvisation, compartmentalization, energy, regeneration, adaptability (as defined by Koshland), or seclusion—guarantee life's presence. They may be necessary for life, but are not necessary *and* sufficient. They are not attributes without which life cannot exist *and in whose presence it invariably does*. Hence, unless something determinative is missing from this list, the materialist's claim of identity between our material incarnation and life is incorrect.

But our materialist might cry foul. "Doesn't this misstate the claim? I do not claim that *each* necessary feature of life is sufficient to guarantee life in its own right. It is only in their community, only when they are all present that life comes into being, and it does so automatically with no intervening process or event. *This* is my claim of identity."

Together, identity and automaticity relieve us of the need to provide explanations for why life comes into being from the mere presence of its various necessary features. Life is simply the workings of the body's own necessary parts and features in their totality. In their community, a mechanical transition occurs from the necessary to the necessary and sufficient. There is no need to explain what new lifegiving feature arises from their convergence—none does. Life simply comes into being.

This insouciance is based on an analogy to chemical reactions. In chemical reactions, all that is required for the end products to come into being is the presence of the necessary reactants, the substances that produce them. Though individual reactants are unable to cause the reaction and produce its products, when they are all present, they become sufficient, and the chemical end products emerge automatically, mechanically. Nothing else is needed. Nothing need be added. The material elements of the reaction themselves entail, that is, cause everything.

In the analogy to life, the necessary material features of living things are like reactants. Though unable to produce life separately, when present together they do so automatically. They entail; they cause everything that is life, and as such, life's sufficiency arises from its necessary causes. This is a very powerful claim because it makes defining the nature of aliveness moot. Our inability to define it is not really a problem. Beyond specifying its necessary features, no definition is needed. Aliveness is simply the result of the presence of life's necessary material antecedents, whatever they may be.

Since doubtless chemical processes are central to living things, many being critically necessary, and despite how hard they are to specify in their totality, it seems eminently reasonable to imagine that life is akin to some kind of extremely complex chemical reaction that displays the same automaticity as chemistry, except that rather than the end product being a chemical, it is life itself.

However effortless, this is an elision. The analogy to chemistry does not merely claim that life and chemical reactions are similar in this or that respect, like a metaphor. What is claimed is that they are analogous in all meaningful respects. The claim, though I confess that I have not seen it explicitly stated, is of an identity between the two in the same way that if $a \times b = c \times d$, then the ratios $a/c$ and $b/d$ are analogous. If the analogy were merely an inference from the similarity of two things in regard to one or several of their properties to all of them, then as with a metaphor, the inference, however suggestive, would prove nothing.

An identity theory of life requires that the two—life and its chemical reactions—be similar in all respects and in a most general sense, and this claim must be proven. It must be shown that all of the properties of the two things are indeed equivalent, or analogous. *None* can be dissimilar. This means, for example, that we would have to be able to explain human psychology and culture in terms that are formally equivalent to chemical reactions.

The automaticity of chemical reactions, their intrinsic entailment, has been proven countless times in rigorous kinetic and thermodynamic terms both experimentally and theoretically. If the analogy to life applies, then it requires the same sort of proof. We must show in similar rigorous kinetic and thermodynamic terms how life's various necessary particularities give rise to life itself.

When all is said and done, the materialist perspective does not excuse us from having to explain life's nature beyond cataloguing its constituents and their interactions. Like all other claims, to be a matter of science, it must be validated. That is, it is not enough to merely proclaim an identity between life and chemistry, between life and its material incarnation, it must be proven. Depending on the assumptions, this proof can have one or more of the following three forms. We must:

- ▸ Discover or articulate the lifegiving feature that emerges from the various necessary material elements of living things,
- ▸ Explain what about the coming together of these elements makes the object alive otherwise, or finally
- ▸ Produce equations and experimental data that show the kinetic and thermodynamic identity of life's necessary elements with life itself, as in chemical reactions.

I have focused on a *chemical* analogy here because it is the most obvious and widely imagined, but similar analogies with the same requirements for proof can be envisioned for any automatic process of our choosing—for example, for physical forces like diffusion or adsorption.

In any event, even with an enormous fund of information at our disposal, the materialist's claim of identity between the chemical and physical features of life and life itself has yet to be realized, and as we shall see shortly, it is very unlikely that it can. Indeed, there is nothing to recommend it beyond the claim itself, and as said, claims alone, however convincingly or forcefully made, will not do. And yet this said, ironically the materialist perspective is often thought to be rigorous and scientific, presumed true even in the absence of evidence, while criticism of it is thought to border on the magical.

The reason for this peculiar state of mind is that since life's underlying chemical and physical basis can be rigorously understood, the materialist's assertion of identity, likewise seems scientifically rigorous, an expression of Ockham's razor, science's rule of parsimony. But this is unscientific hogwash at the least and sophistry at the worst, and in any case it is a terrible misuse of analogy. We can no more say without proving it that life automatically arises from its material necessity than we can say that because bats fly and birds fly, that bats are birds, or that if I am tall, I am a basketball player.

Now we should be able to understand why Koshland's scientists thought they knew what life was and could list its central features, but nonetheless found themselves unable to define it. They were powerless because they were incapable of showing that various well-understood, necessary features of life were in themselves sufficient to produce it, to produce aliveness. To understand what makes something alive, to discover this deepest aspect of life's nature, to establish the character of the phenomenon as a matter of science, it is necessary to identify and understand the sources of life's sufficiency in their own right. This is simply because only in their presence do we find life.

Materialists have only one remedy for their inability to provide affirmative evidence for the notion that life appears automatically in the presence of its necessary elements, for the automatic conversion from mere necessity to sufficiency. They must prove that there is no possible source of life's sufficiency other than its material antecedents. But such proof is impossible. All that can be done is to claim provenance by exclusion, that is, by saying that there is no other way it can happen. This is the fallacy of the false alternative—just because we cannot imagine something does not mean that it does not exist. But fallacy or not, as we will see in the next chapter, not only can life's sufficient causes be identified, and identified elsewhere, the materialist's claim of identity is false and rather obviously so.

# CHAPTER 5

# A Sufficient Property of Life

## A Search for the Properties That Attest to Our Being Alive

The mechanisms of life can be unveiled and proved only by
knowledge of the mechanisms of death.

**—CLAUDE BERNARD, *AN INTRODUCTION TO THE STUDY
OF EXPERIMENTAL MEDICINE,* 1865**

THE MATERIALIST claim of identity is refuted and life's sufficient properties uncovered, not in the circumstances of life, but
in those of death, and this is not a new, but an age-old realization. In ancient times, clues about the cause of death were
in great part limited to surface properties—what one could see,
smell, feel, and hear. The dead person no longer moved, no longer breathed, had no heartbeat, and became pale and eventually
cold and hard. Consequently, it seemed that whatever had generated movement and breath, made the heart beat, and gave color,
warmth, and flexibility was gone. Some lifegiving property had
been lost; some essence had left the now motionless body, and
yet—and here is the point—the body otherwise seemed materi-
ally unchanged. It looked much the same just after death as when
the person was alive.

Today when we can examine the microscopic features of the
body, the underlying anatomy and chemistry of cells and tissues,
we find that like the body's gross features, things look much the
same immediately after death as before. Cells contain the same

structures in the same places. There is still DNA and its program. Thousands upon thousands of protein molecules are in place, unchanged, acting, or if not acting, available for action, and yet the organism is no longer alive.

As even our most primitive human ancestors appreciated, death leaves behind our material incarnation as it takes life. In the language of necessary and sufficient properties, our necessary features remain in place despite being abandoned by life, while our sufficient properties are irrevocably lost, disappear one and all, and represent that abandonment. Death, it turns out, is their loss. This separation of life's sufficient properties from its necessary ones shows simply and definitively that the identity theory of the materialist is false.

## Life as Extrapolation

In this case, the task confronting us is to identify the authentic source of these lost sufficient properties. If they are not to be found in our material embodiment, then where are they found? Before we travel this path, we must deal with another notion of physico-chemical identity, a consequence of cell theory that has been with us for close to two centuries. In it the organism is seen as a compilation of other living things—in particular, cells. As such, cells are the material objects that give rise to life. By forming tissues and organs, and systems of tissues and organs, they give rise *ipso facto* to whole living organisms. Accordingly, living cells in their totality and the living organism are one and the same thing.

But this is a serious misperception. Cells, even groups of cells organized into tissues and organs, assuming them to be autonomous living bodies, are not sufficient to produce life in the whole organism. Indeed, in most cases upon death, the cells not only remain in place, but functioning, perhaps only fleetingly, for seconds, minutes, or hours, but functioning nonetheless. That is to say, we can have living cells and a dead body. As such, life has an existence beyond its constituent cells. The two are not automatically linked. Though cells give rise to all of life's incarnations, they are not sufficient to guarantee life in the organisms they form. Even in their entirety they are not a sufficient property of life in multicellular organisms.

## The Search for Sufficiency

If we cannot look to chemistry, physics, or even living cells for the cause of life's sufficiency, then where can we look? What is it that is lost upon death? The modern pathologist can describe death's various causes in humans with great specificity. For example, there may have been a coronary thrombosis that damaged heart muscle (an infarct), a burst aneurysm, cancer in a vital organ, or perhaps the patient was infected with bacteria that produced a poisonous toxin.

As it turns out, underlying each of these, indeed underlying all causes of death in humans, is a common insufficiency—*the inadequate perfusion of blood*. It is the ultimate cause of death, whatever its generating or penultimate cause may be. For example, a physician might tell us that a thrombosis (clot) impaired the coronary circulation (the circulation to the heart) and consequently damaged heart muscle, leaving the heart unable to pump blood at a great enough pressure to adequately perfuse vital tissues and organs, including itself, and that this resulted in a deficiency of oxygen at crucial sites around the body, which led inevitably and quickly to death.

By saying that perfusion was inadequate, I mean that the *flow* of blood through the vasculature, through the circulatory system, usually expressed in milliliters per minute, had been reduced to the point that oxygen was not being provided to the person's various cells and tissues at a sufficient rate. There are three potential causes for such a reduction. The one just mentioned, a reduction in blood pressure—the force that drives blood through our vessels—as in a heart attack, and two others.

The second cause is a reduction in the *amount* or volume of blood that is available for perfusion, as occurs after a hemorrhage. A burst aneurysm produces an internal hemorrhage in which blood leaves the circulatory system and enters the interstitial spaces of the body (the spaces between cells and the circulatory system). As a consequence, the amount of blood in the vasculature is reduced, and consequently so is blood flow. In this case, it is reduced blood volume, not damaged heart muscle that is the problem. But just as with a heart attack, it results in essential tissues and organs being inadequately perfused and oxygen deficient. Since the inadequate perfusion includes the heart itself, a secondary heart attack may supervene, further reducing the ability of the heart to circulate blood. Of course, this can also happen if the hemorrhage is external, such as a knife-wielding felon putting a large gash in your chest, piercing major blood vessels in the process.

The third cause of inadequate perfusion is a large change in the *resistance* offered by the blood vessels to fluid flow. This may be due to an *increase* in resistance caused by an "occlusion" or narrowing of blood vessels, as in atherosclerosis, making it harder for blood to get through. Or conversely, and paradoxically, it may be due to a *decrease* in resistance, with blood vessels "dilating" or opening up. In this case, blood pools in the large veins of the body rather than being returned to the heart for recirculation. This is what takes place in congestive heart failure. Increased resistance by reducing blood flow to the coronary circulation can produce a heart attack and damage heart muscle, diminishing, as said, its ability to pump blood. Whereas the pooling of blood in the large veins of the body reduces the *effective* blood volume, that is, the amount available for circulation, just like a hemorrhage.

But even if death's proximate cause lay outside the circulatory system, such as an infection or cancer, inadequate vascular perfusion remains its ultimate cause. For example, in cholera, the *Vibrio cholerae* bacterium secretes a toxin in the intestines that causes severe diarrhea and dehydration. The loss of fluid caused by diarrhea reduces the volume of blood. This in turn leads to a reduction in blood flow, the inadequate perfusion of vital organs with oxygen, and consequently to death.

## A Sufficient Cause

And so, whatever its initial cause, death is ultimately the result of cardiovascular collapse, leaving the body unable to adequately perfuse its vital tissues and organs with oxygen-containing blood. *It is in this failure that a sufficient cause of life is lost.* Vascular perfusion is a *necessary* condition for life in species like ours because it is the means the body uses to get oxygen to our cells and tissues, but it is also a *sufficient* condition. However impaired we may be in other ways, as long as our organs and tissues are being adequately perfused with oxygen-containing blood, we are alive. Brain death notwithstanding, only when vascular perfusion becomes insufficient does life, however limited, end.

There is an important distinction to be made here. We might well ask whether oxygen deprivation, not inadequate vascular perfusion, is the ultimate cause of death. Isn't death really due to *its* absence, an absence that inadequate perfusion merely brings about? Isn't oxygen deprivation the real culprit, the lost sufficient property? If you think about it, you will realize that this does not make any sense.

Oxygen is, of course, a chemical element like any other on the periodic table. If anything is inanimate, it is oxygen. Self-evidently, life is not present whenever oxygen is present, and as such, it cannot be a sufficient property of life. It is not even necessary for life in green plants, the source of the oxygen we so desperately need, which live quite nicely, photosynthesizing in its absence. Certainly, oxygen is necessary for life in humans and most animal species, but it cannot be a sufficient property of life.

No doubt there was plentiful oxygen in the air around the dying individual. The patient may have even been mechanically ventilated with "pure" (95 percent) oxygen, rather than the 21 percent found in air, to increase its availability, but still died. The problem was not the lack of oxygen, but the powerlessness of the cells of the body to gain access to it. However plentiful, it could not be distributed to the body's tissues and organs in adequate amounts due to the decrease in blood flow. Whether the result of the failure of heart muscle (decreased blood pressure), an insufficient volume of blood to circulate (because of hemorrhage or dehydration), or a change in the resistance of blood vessels to flow, its diminution was the cause of death.

But if our goal is to generalize about life's sufficient causes, this is hardly a gratifying conclusion. Though vascular perfusion may be a necessary and sufficient property of life in animals with a circulatory system, with blood vessels, a heart pump, and the like, such species only represent a tiny subset of living things. Protozoa, bacteria, fungi, as well as plants and many invertebrate animals lack this arrangement and yet are undoubtedly alive.

Therefore, vascular perfusion seems a highly unsuitable property for making broad generalizations about life's nature. Indeed, to the contrary, its sufficiency in humans and other species with a circulatory system seems to teach that while a particular property may be necessary and sufficient for life in one type of organism, in one species or group of species, it may be totally irrelevant in another. This is the opposite of a generalization and again brings us back to Daniel Koshland's scientists and their determination that there is no general definition of life, no single life-defining property. If vascular perfusion is critical and life-defining for us, but irrelevant for other life-forms, then the same must be true of other features in other living things—a sufficient property of life here, but irrelevant there. It seems that rather than there being a single life-defining property that applies to all living things, there are different ones for different types of organisms.

## The Way We Are

This is not only a very troubling conclusion for biologists seeking to generalize about life's causes, but for science more broadly. Even though a central goal of science is to explain natural phenomena in the most general and inclusive terms, it seems that as a *matter of science* we cannot say in what fundamental, what general way, living things as a class differ from previously living or inanimate objects. This organism may differ in this way, and that in another. We seem unable to spell out with stipulations that apply to all why some physico-chemical systems are alive and others are not. We cannot specify what common, what general, what fundamental feature bestows life on certain material systems.

What a horrid conclusion. But is it true? Are we really unable to provide a satisfactory general definition of life? Are we really unable to identify a fundamental life-bestowing property (or group of properties) that applies to all living things? Is it as Koshland's scientists determined, that we have no choice in the matter? Like it or not, this is reality and there is no way around it. Does science fail at this critical juncture? Or does this conclusion simply reflect our lack of discernment? Have we yet to uncover, discover, or conceptualize life's essential nature, or as I believe, given voice to what we already know?

What is clear is that if there is some general definition of life, some unique life-defining property that has yet to be articulated, we should be able to understand vascular perfusion—a necessary and sufficient property of life in species like us—in *its* terms. We should be able to conceptualize vascular perfusion in a way that is entirely general and inclusive, that can be applied to all living things, as well as all causes of death. If we can, we will have defined life, exposed its essential property. If we cannot, then we will have confirmed that no such definition is possible.

To make such a determination, we have to look more deeply and more abstractly at vascular perfusion. Let us begin with a question. Why did nature develop a circulatory system in the first place? Couldn't other means have been found to get needed oxygen to our cells and tissues? Why was this particular instrumentality chosen?

The answer is instructive. The evolution of the circulatory system, or something quite like it, was unavoidable *if* human life, indeed if vertebrate life more generally, was to come into being.

To understand why this is so, we need to ask another question. Why didn't nature just leave well enough alone? Why weren't the forces of natural selection satisfied with the first world they created—a world filled with single-celled organisms?

Why didn't things just stay that way? Why did animals of the size and complexity of humans come into being? What moved nature down this path? Their creation, our creation, certainly does not seem inevitable. Religious belief aside, humans, mammals, and vertebrates in general have no special claim to existence. Why did they evolve? Was it merely a matter of chance as is often thought, or was something else involved?

## Evolution's Strategies

To answer this question, we need to consider the world of single-celled organisms—not the organisms themselves, but the world in which they live. The most significant fact about that world is that it is incredibly small. Even a most powerful protozoan able to move very rapidly through water on its own is restricted to a "geosphere" not much larger than a drop of water—an infinitesimally small fraction of the planet's surface, a truly Lilliputian environment. The organism's survival depends on the benevolence of this tiny world in the same way that ours depends on that of the earth at large. It can no more escape it than we can escape the earth (actually, given human space travel, less easily).

It turns out that on this small size scale, massive changes, major environmental shifts, and cataclysms occur far, far more frequently than on a planetary level. Consequently, single-celled organisms face repeated, if not incessant, extremes in their environment—for example, changes due to the mass flow of water or soil on a microscopic scale when it rains or the wind blows (or more personally when we human Gullivers splash water or step on wet earth). Alternatively, a microscopic Garden of Eden might quickly ebb away as necessary nutrients are used up, altered, or destroyed, or commandeered by competitive organisms. Early life not only faced such catastrophes with great frequency, it did so ill prepared, passive, and exceedingly vulnerable to nature's whims.

To overcome this perilous and precarious state of affairs, nature allowed and evolution took advantage of three strategies. Following one or another, life's foundational adaptations to environmental circumstance arose. Using the first, organisms remained passive, taking what they were given, going where they were bidden, and hoping for the best. Through the wonders of natural selection, species that followed this path came to possess a wealth of remarkable coping skills despite their passive nature. The major evolutionary products of this strategy are today's protozoa and bacteria. Bacteria in particular are the most flexible of creatures, able to survive all kinds of dire circumstances. Indeed, it is their flexibility that makes the treatment of infectious diseases a never-ending task for medicine. Give a particular antibiotic, and natural selection favors the few pathogenic organisms that have ways to avoid its toxic consequences. Take a needed food source away from a colony of bacteria, and some cells will survive on whatever remains; or take away all usable food, and desiccated spores will form to wait for a better day. They are passive but persevering.

With the second strategy, the organism attempts to gain some control over its situation. It tries to prevent itself from being moved hither and yon at the mercy of catastrophes in its microcosm by staying put. Communities of cells literally put down roots and became "sessile," immovable. Like organisms that followed the first strategy, they developed adaptations of passive resistance to overcome less welcoming conditions when they became manifest. Many marine invertebrates, from sponges to sea anemones, and most plant life, fungi, and some bacteria are products of this strategy. Drought resistance in plants is a good example of an adaptation to this lifestyle.

But it is the third strategy that concerns us here. In this case, the organism attempts to gain control over its situation, not by staying put, but by moving from one environmental niche to another. As said, though many single-celled organisms move very quickly on a microscopic scale, given their size and the energy available to them, they just cannot move far enough or fast enough to escape the tiny world in which they are embedded. With the help of natural selection, some organisms found ways to break out of this small world, to move over macroscopic and eventually geographic distances. Over hundreds of millions of years, animals, most conspicuously vertebrates, evolved this capability. They were able to leave meager or unsafe environments to seek sheltered and nurturing havens elsewhere.

So, take what you are given, stay put, or move. These are the three choices that nature provides and that defined the paths along which living things evolved. Natural selection helped bacteria and protozoa evolve in ways that allowed them to adjust effectively to environmental challenges given their physically passive nature (by "passive" I mean the inability to move on their own beyond microscopic distances). Most plants and many marine invertebrates have become masters at the art of staying put. And finally, animals of all sorts move as far and wide as their physiology allows, from mere inches to circling the planet, ever seeking greener and safer pastures.

Before moving on, I should be clear that not only the "movers" are able to circumnavigate the globe. For passive or sessile organisms, movement is accomplished not by their active contrivance, but on the winds and waters of chance. For example, many bacteria, protozoa, single-celled plants, as well as simple multicellular organisms, such as jellyfish, are spread across the face of the earth by the flow of water, or on or within various animal species. Sessile plants are dispersed widely as seeds, not mature organisms, on the wind (as well as on the bodies of animals).

## The Size of Life

But only certain animal species migrate as a matter of active preference, by something akin to what we call "free will." To do this, they had to be able to negotiate macroscopic geographic distances in reasonable periods of time. And in turn, to do this, first and most obviously they had to be far larger than their microscopic ancestors. Though stamina (energy) and speed (mechanics) are important, the distance an organism is capable of moving is quite simply proportional to its size.

Even though both tiny birds and gigantic whales can migrate over long distances, and conversely, many large animals stay in circumscribed geographical areas, if we move a ten-micron-long body—about the size of an ordinary biological cell—one body length, we will have moved it ten microns, whereas if we move a one-hundred-cm-long body (about three feet) one body length, we will have moved it a million microns—one hundred thousand times farther. Given energetic and mechanical constraints, only by increases in size could organisms propel themselves over significant geographical distances in a timely way.

Still, size does not produce movement. Mechanisms are needed to animate a large body, just like a small one. Building on the means used to produce movement in single-celled protozoa, a mechanism we call "contraction" evolved that is responsible for essentially all macroscopic movement in animals. Certain cells—muscle cells—became specialized to contract or shorten. This occurs when within the elongated muscle cell, two different filaments—one made predominantly of the protein actin and the other of the protein myosin—slide past each other along the cell's long axis, shortening it in the process.

This movement is driven by chemical energy released from the breakdown of life's premiere energy molecule—adenosine triphosphate or ATP—at sites called cross-bridges that connect the two filaments. The chemical energy in ATP is converted to the mechanical energy needed to push the filaments past each other, the cross-bridges acting like ratchets. Cells in muscles are organized end to end, or in series, as well as side by side, so that the microscopic sliding of molecular filaments that shortens a given cell adds to the shortening of many others, which taken together, produce a macroscopic shortening or contraction of the whole muscle.

## Muscle and Bone

But muscle is not enough. Certain animals, like snails, that depend solely on muscle contraction for movement crawl slowly across the earth's surface, alternately contracting and relaxing their ambulatory muscles. But of course they move at, well, a snail's pace. Though they live in a far larger world than protozoa, snails and other species that depend on muscle alone can only move about on a small corner of the earth's surface by their own means.

To achieve long-range movement, much less movement with a planetary reach, something was needed that muscle contraction alone could not provide. Once again through the agency of natural selection, nature found ways to amplify the action of muscle by incorporating a mechanism with a much greater mechanical advantage. Appendages suited for walking, swimming, or flying evolved, and depending upon the medium in which they lived, animals came to use them to crawl, walk, or run on or in the earth, swim in its seas, or fly in the air above.

When we contract the muscles attached to our limb or long bones, they become tense, but initially not significantly shorter. This is because they are attached to rigid bone at both ends and can't shorten (in this case, the sliding of filaments within muscle cells is not immediately converted to the kinetic energy of shortening—this is called "isometric contraction"). Converting the tension that is produced by contraction to the mechanical action of shortening is the job of joints, which is accomplished by the fact that each muscle is attached not to one, but two bones straddling the joint between them. Contraction produces movement of the bones around the joint, allowing the muscle to shorten and the organism to move.

It is in this way that the kinetic energy generated by muscle contraction produces long-range movement. What a remarkable result this is! In its creative genius, evolution contrived a way to literally "leverage" the microscopic movement of filaments within tiny muscle cells to carry organisms across the face of the planet. Imagine a bird that lacks wing bones but is still somehow able to fly when its flight muscles contract. If these muscles averaged ten centimeters in length and shortened in half with each flight stroke (reasonable values), the boneless bird could (though unlikely) move half the composite length of its flight muscles. A few hundred contractions would move it a few yards at most. But the same number of contractions of the same muscles attached to wing bones might carry it miles.

But even muscle, bones, and joints are not enough. To produce effective running, swimming, or flying, muscle contraction and the movement of limbs must occur in a particular sequence. Without coordination, movement is difficult, if not impossible, as in various spastic muscle diseases. To achieve efficient movement, a complex web of nerve cells signal the various muscles to contract in a sequence proscribed by the central nervous system.

Moreover, as evolution proceeded and as muscle mass became greater, more food was needed to provide energy for contraction. This required an increase in the organism's ability to obtain and process food. Our complex gastrointestinal system, with its large glands, its various tubes and cavities, as well as a multitude of activities designed to more effectively seek, capture, and process food was born of this need.

In addition, mechanisms were needed to excrete large quantities of waste product (e.g., fluid by the kidneys and gas by the lungs), to coordinate all of the various complex activities of the large body (the nervous and endocrine systems), as well as to regulate body heat and water. And though muscle remained the largest tissue, to fulfill these coexisting needs, substantial mass was added to the body. Although the size of animals varies greatly, to move long distances required a larger facility. Like a growing corporation, a larger building with more space to accommodate expanding and increasingly diverse needs was essential.

## Why the Cardiovascular System?

All well and good, but why does any of this require a heart and blood vessels? Why does it necessitate vascular perfusion? Why does an increase in the size of organisms demand a cardiovascular system with its ability to perfuse tissues and organs with blood?

To understand why, we need to ask another question. More cells, not larger cells, produced proportionately larger organisms. But why—why not larger cells? Why was the increase in the size of organisms that evolution produced the result of larger and larger agglomerations of cells, not increases in cell size? Couldn't the advantages of size have been incorporated into a cell as large as, indeed congruent with, the size of the organism? In fact, wouldn't this have been a simpler way for evolution to proceed? Why were multicellular organisms needed, or were they?

Cell size increased by two to three orders of magnitude during evolution, from cells roughly as small as today's bacteria ($4$ $\mu^3$) to the average size of the modern eukaryotic (nucleus-containing) cell ($4 \times 10^{2.3} \mu^3$). Though this is a large increase, the difference between the volume of a small bacterium and a hundred-foot-long whale is not two or three orders of magnitude, but about twenty-two orders of magnitude ($10^{22}$)! That is, only 0.0000000000000000001 percent of the difference between the size of a bacterium and a whale is attributable to increases in cell size.

Of course, the difference is much less for smaller animals (knock off a few decimal points for a man or a rat), but whatever sized creature we consider, increases in the mass of animals, in the mass of all multicellular, all metazoan organisms has been almost completely the result of increases, often titanic, in cell number. None of these creatures, not even those as small as the tiniest insects or the smallest round worms, are comprised of a single giant cell. But why not—was it just a matter of chance or was it fated? As it turns out, it was inescapable. Nature, physical law, demanded it. If the size of living things was to increase from the microscopic to the macroscopic realm, then more cells, not larger cells, were needed. There were no ifs, ands, or buts about it.

## Flow and Diffusion

We can understand why by looking at the way molecules move. There are two. They can be carried along in the medium in which they are dissolved or suspended as it moves. We call this "flow," "mass flow," or "mass transport." When water and all of the substances suspended and dissolved in it are carried out of the mountains into the valleys and toward the sea, it is, of course, due to the *flow* of water. As we have discussed, the movement of blood through our arteries and veins is likewise the consequence of flow. In both cases, motion is generated by a hydrodynamic (fluid) pressure gradient or pressure difference. For rivers, the difference is one of altitude between the mountains and sea. For the circulatory system, it is the pressure difference produced by the contraction of heart muscle.

The other way that molecules move is inherent to them. When dissolved in a medium like a liquid, they move randomly due to their own intrinsic propensities, their own energy. This is known as Brownian motion and it gives rise to *diffusion*. In diffusion, dissolved "solute" (nonfluid) molecules collide with molecules of the medium in which they are suspended (the solvent) in an elastic interaction that propels them. When there is a high concentration of a particular molecule at a location, collisions occur more frequently, moving them farther and farther away from their source toward areas where they are sparse. The rate at which this occurs depends on the difference between the substance's concentration at its source and its destination, as well as the physical properties of the molecule being moved and the medium through which it is being moved. This relationship was formalized in the nineteenth century in an equation called the Fick diffusion equation, named after the scientist who first described these relationships.

At first glance, of the two transport modalities, diffusion seems far more desirable. Unlike rivers, flow in living things usually requires special, usually complex, external mechanisms, *deus ex machina*, like the heart and the circulatory system. Diffusion, on the other hand, requires none of this—no external machinery, no pumps, no plumbing. Molecules just move on their own accord, providing both the energy and the mechanism themselves.

For example, oxygen would simply travel down its concentration gradient from the air to the site where it is needed. It would enter our bodies across the skin, dissolve in the liquid that comprises the skin, and then diffuse throughout the body's fluids from areas of high concentration at the surface to those of low concentration at the core, supplying oxygen along the way.

Why didn't nature choose this simple mechanism to provide us with oxygen? Why did it choose the complexities of the circulatory system, with its arteries, capillaries, and veins, its chambered muscular heart, and the nervous and hormonal mechanisms needed to control blood pressure and flow? Why did natural selection favor this far more complex approach when a simple physical method was available? Isn't parsimony nature's rule?

For all its simple charm, diffusion presented a serious, indeed an insurmountable, problem for moving molecules like oxygen into organisms *if there was to be a major increase in their size.* While diffusion is very efficient for movement over short distances, a few microns or less, such as into and across small biological cells, it becomes rapidly (exponentially) and profoundly inefficient if the distance to be traveled lengthens even a little bit, say across a cell two or three times the average size of cells in eukaryotes, no less across a massive cell equal to the size of even a small metazoan. To give an extreme nonbiological example, if nature had to depend on diffusion to move substances in rivers from the mountains to the sea, they would never get there, at least not in the lifetime of our planet.

The Fick equation tells us that we can only depend on diffusion when the distance to be traveled is *microscopic.* As such, the diffusion of oxygen across a cell membrane—about 0.01 micron thick—takes only a small fraction of a second, while its passage across a human hair—about fifty microns (about 1/80th of an inch)—requires close to a minute. So for diffusion to work, our cells had to be small. Nature had ordained it. But even so, to diffuse from our body's surface to its center, say twenty centimeters (about eight inches), would take many hours however small the cells being traversed, and we just cannot survive the wait.

If we could magically delete the circulatory system with the press of a button on a great computer, though oxygen would continue to diffuse into the body across its surface, we would die almost immediately. Diffusion offers too little, too late. Even if air were 100 percent oxygen, not 21 percent, it still wouldn't work. The only way to effectively oxygenate bodies as large as ours in this fashion is in a world where atmospheric gas pressures and oxygen concentrations are far greater than found naturally on earth (as in hyperbaric surgery). For the circumstances of our planet, if we only had diffusion to depend on, animals could not be much larger than the thickness of a human hair.

That is, without mechanisms to move substances by pressure-driven fluid flow, animal life beyond the materials, processes, and events that can be accommodated within single microscopic cells or thin layers of cells would not have been possible. To achieve the advantages of size, structures and mechanisms had to evolve that could carry an oxygen-containing fluid throughout relatively large bodies comprised of multitudes of cells in a matter of seconds, not hours. In an evolutionary process that took hundreds of millions of years, natural selection produced a system that does just that. In its blind brilliance, it produced the heart and its muscle to propel fluid, as well as the conveyances—arteries, veins, and capillaries—needed to distribute it. If animals as large as humans were to come into being, if they were ever to exist, the circulatory system was not an option. It was not a matter of chance or choice, but an absolute necessity. Life would have been a far more limited, as well as a far less interesting and less remarkable phenomenon, if substances could only move by diffusion.

## The Sought-after "General Phenomenon"

And so, vascular perfusion was not merely useful, it was an absolute necessity for large animals. But what of the broad generalization we have been seeking about life's sufficient causes? How does any of this enlighten us? In what way, if any, can vascular perfusion be said to apply to all living things, to all creatures, large and small, even those that lack such a system? Beyond describing the particular material incarnation that allows for perfusion in large animals or giving details of the physics of fluid flow that produces it, can anything be said more broadly about vascular perfusion?

Can it be thought of in terms of some all-encompassing generalization? Though it may seem odd, it can. To see how, we have to think not of the material substance and form of the circulatory system, nor the physics of fluid flow, but the *function* of vascular perfusion, its underlying purpose. Quite simply, as explained, along with several other mechanisms (see chapter 7), vascular perfusion is necessary for large animals to obtain adequate supplies of oxygen *given its concentration in the atmosphere.* Without it, they would not get enough and consequently could not survive.

What this says is that vascular perfusion is an *adaptation* to a particular circumstance of our environment—the concentration of the gas oxygen in the earth's atmosphere. It is this, its adaptive quality, that it holds in common with a multitude of other mechanisms and processes with wholly different incarnations and purposes. This is the generalization that we have been seeking. *Vascular perfusion, a sufficient property of life in creatures like us, is an adaptation to environmental circumstance!* As we shall see in what follows, whatever their particular purpose, whatever their specific incarnation, all of life's sufficient properties—properties like vascular perfusion that indicate life's presence—are or are related to adaptations to environmental conditions.

CHAPTER 6

# Dr. Bernard's Adaptations

## *The Internal Adaptations of Life and Their Critical Role in Understanding Life's Nature*

———— ⌒ ————

> [The living being must be stable] in order not to be destroyed, dissolved,
> or disintegrated by the colossal forces, often adverse, which surround it.
> By an apparent contradiction it maintains its stability only if it is
> excitable and capable of modifying itself according to external
> stimuli and adjusting its response to the stimulation.
> —CHARLES RICHET, NINETEENTH-CENTURY
> FRENCH PHYSIOLOGIST

REMEMBER THAT a biological adaptation is any feature of an organism that improves its situation in the face of a particular environmental challenge or hazard. And though we commonly think of the concentration of oxygen in air as an environmental *constraint*, not a challenge, and physical laws such as the Fick equation as *properties of nature*, not hazards, the difference is just semantic. They are environmental challenges and hazards if anything is. If the constraint is not accommodated, if the law is not complied with, the organism cannot survive. Consequently, like so many other properties of living things, vascular perfusion is an adaptation to environmental circumstance, and a critical adaptation at that.

But for those of you who have read about evolution in a textbook or in the many popular books on the subject, or if you have actually tackled *On the Origin of Species* itself, vascular perfusion may not jump to mind as a prominent example of a biological adaptation. Indeed, what we commonly think of as biological adaptations seem to be different kinds of phenomena altogether. Perhaps the many shapes of the beaks of the famous Galapagos finches, each suited to feed efficiently in a particular microhabitat, come to mind. Or perhaps you envision the color of things—an animal blending in with the earth, the bark of a tree, or a leaf to hide from predators, or a colorful flower attracting pollinating bees. Or possibly you see a speeding gazelle escaping an attacking lion, or the claws and teeth of the lion at work after it catches its prey. Scientists have described a host of these phenomena, sometimes obvious, sometimes arcane, in varied and sundry species for equally varied and sundry situations. As said in the introduction, "From this external perspective, adaptations are understood to be the skills necessary to obtain food, to defend against becoming food for others, to endure extremes in temperature and weather more generally, and critically, to procure sex—all in the service of survival and propagation." Adaptive properties are usually imagined in these terms—that is, in terms of features that can be observed by watching organisms from the outside as they live their lives (or that can otherwise be inferred from their surface appearance or remains). These are the phenomena of Darwin and the naturalists, and the focus on them has in great part been due to the fact that, like Darwin himself, most biologists who have evinced an interest in evolution and biological adaptations have been trained to observe and analyze nature in this fashion.

But no matter how carefully or expansively we look at life in this way do we learn much if anything about vascular perfusion. Indeed, we might not even realize that it exists. To learn of its presence, to understand its nature, we have to look *inside* opaque living organisms. That is, vivisection—carrying out experiments on living animals—that oft-derided enterprise, is essential to identify, much less understand, vascular perfusion, and for that matter almost everything else about our physiology.

## Abstaining from the Struggle for Survival

In addition to definitional confusion between physical constraints and nature's laws on the one hand and environmental challenges and hazards on the other, this historical circumstance explains why vascular perfusion—by my reckoning, a critical biological adaptation—is missing from customary lists of these phenomena. But there is another reason, a third reason, a mistaken perception, and it is an important mistake.

As I sit in my reclining chair, alone in a room at a comfortable temperature, my legs elevated, doing absolutely nothing of significance, as mentally disengaged as I can be, I seem removed from the struggle for survival. I am not seeking food or sex, nor am I fending off an attacker. I am not even thinking about such things. I am neither hot nor cold. In this comfortable haven, perhaps dozing off to sleep, I seem relieved, if only for a moment, from engagement in the struggle for survival.

In context and inaction I seem "adaptationless," however varied and advantageous my adaptive properties may be when they are engaged. I am not making use of my adaptations, nor do I feel the need of them. But this is merely an illusion. However at ease I feel, however inactive I may be, however seemingly pacific my surroundings, and however many of my adaptations are *not* being expressed because of my particular circumstances, the perception that I am for the moment an adaptationless living being set apart from the struggle for survival is badly mistaken.

As it happens, however detached I seem, countless adaptations, unseen and unbidden, are hard at work fighting ceaselessly on my behalf. The undeniable fact is that beneath my passive bearing, outside my direct perceptual grasp, there is incredible pandemonium, turmoil without end, as my engagement with the environment, my struggle for survival, goes on. Crucial battles are being fought and innumerable adaptive properties deployed within my resting torso, every second, every minute, every day of my existence.

My lack of awareness of these adaptive properties is not due to inattention. I would be in a dire strait—overwhelmed, confused, and confounded, unable to act, to integrate it all—if I were in touch with even a small percentage of them. Indeed, my lack of awareness is itself a pivotal adaptation that enables me to navigate very complex circumstances without having to consider everything about them all the time.

Though I need to be aware when crossing the street, or am not aware at my own risk, within my body *the internal adaptations of life*, operators in my ongoing struggle for survival, do not require mindfulness. In their multitude and variety, they are rendered automatic, outside my conscious concern. But this makes them no less real and certainly no less important. To the contrary, counted among their number are life's central features, phenomena such as vascular perfusion that work long and hard to ensure my survival.

## Dr. Bernard's Adaptations

I think of the internal adaptations of life as being those of Dr. Bernard, not Dr. Darwin. Perhaps they should be called Bernardian rather than Darwinian adaptations. Certainly Charles Darwin and his supporters were aware of their pivotal role, but it was left to the French physiologist Claude Bernard, a contemporary of Darwin's, a masterful experimentalist and observer, and a committed vivisectionist, to articulate their nature and significance with what we might call modern clarity.

Though his contemporaries Darwin and countryman Pasteur are certainly more famous, Claude Bernard's contributions to science were no less impressive, and that is, of course, saying a great deal. Over a period of some thirty years in his "dark, dank tannery" of a laboratory, as professor of physiology at the Sorbonne, Bernard, his students, and his assistants carried out a remarkable series of animal experiments that made immense contributions to our understanding of how the body works, especially its adaptive properties.

Like Darwin, Bernard was a great observer; like Pasteur, he

was a great experimentalist—the greatest experimental physiologist of his time, perhaps of all time—and like them he had the
intellect and courage to imagine broader meanings to what he
found. His contributions were not merely important, not merely
manifold, not only have they stood the test of time, but some
have served as critical cornerstones for our modern understanding of life.

To take a noteworthy example, Bernard was the first to demonstrate metabolism (what he called "nutrition"). Metabolism is
the sum total of the chemical reactions that transform matter and
produce and use chemical energy in cells. DNA aside, they are
life's chemical basis. In Bernard's words, they are the reactions of
life's chemical "creation" and "destruction." In this work, he discovered that glucose, our primary source of energy (it gives rise
to ATP), was manufactured (glycogenesis) and stored (as glycogen—a polymer of glucose) in the liver. As a consequence of these
findings, he was the first to establish the existence of chemical
creation (of glucose), energy storage (glucose in glycogen), and
biochemical polymers (glycogen) in the body. These discoveries
served as the substrate, the mother liquor, for the later unearthing of our complex metabolic cycles, the elucidation of which
became the major occupation of biochemists in the twentieth
century, starting with the work of the great biochemists Meyer-
hoff and Krebs in the same liver tissue. Also, as part of this work,
Bernard discovered the chemical basis for the digestion of food
(along with its major locale—the intestines, not the stomach as
had been thought) and in so doing provided the first evidence (in
the pancreas) for the molecules we know as proteins (in particular, the digestive enzymes).

No minor accomplishments these! Though this certainly would
seem enough to have secured an important place for Claude Bernard in the history of science, as it happens, this is only the beginning of his story. These impressive contributions were matched if
not surpassed by another—his discovery of the mechanisms that
are responsible for *regulating* life's central internal adaptations

in complex animal species. Bernard discovered what determines how much blood the heart pumps, at what pressure and frequency, what determines the rate of blood flow through our various blood vessels and its distribution among our diverse tissues and organs, as well as the mechanisms that fix how much air we breathe and the composition of the gas we expire.

To this we can add the means the body uses to manage the digestive process—everything from the rate at which food is propelled down the gastrointestinal tract, to the rate of secretion of the enzymes responsible for digestion, to how much food we eat. Add further the mechanisms that play a critical role in regulating the excretion of waste products, the maintenance of proper hydration and temperature, and even the performance of sexual acts. And if this is not enough, now include the processes that motivate us to change our circumstances. Do we need to cool down or heat up, do we need to drink water or refrain from drinking water, should we attack or flee, and should we seek sex or food? Quite an impressive list indeed!

Bernard discovered that regulating these adaptive mechanisms and more is the responsibility of the autonomic nervous system, a part of the nervous system that resides in the body's parenchyma, not its spinal cord. We call it the *autonomic* nervous system because it functions autonomously, outside our conscious direction. It has two branches—sympathetic and parasympathetic—that usually act as polar opposites—one serves to increase a particular activity while the other reduces it.

The primary task of the parasympathetic branch is to enhance "vegetative" activities—those concerned with digesting and assimilating food. When we say that we are "vegging out"—relaxing after a good meal, reading a book, watching TV, or just taking a nap—the parasympathetic nervous system is in the ascendance. On the other hand, when we are concerned with *obtaining* food, *not becoming* food for others, protecting ourselves against any and all perceived environmental threats, or are *seeking* sex, the sympathetic branch dominates. In all, it energizes, arouses, makes us anxious, and enables us to "fight or flee." It empowers us to actively engage our environmental circumstances, rather than passively accepting them. Sex, not surprisingly, is a confusing matter. For example, in the male, erection is a parasympathetic (vegetative) function, while ejaculation is a sympathetic (energetic) one.

In any event, sympathetic and parasympathetic inputs act simultaneously and are balanced against each other to produce a particular maintained, most salutary, most adaptive state. As a consequence, we have, for example, a "normal" heart rate that is characteristic of us as individuals and of humans as a species. More correctly, we have a range of normal rates that vary for different degrees of activity and different needs. This impulse for "normality" holds for all parameters under the direction of the autonomic nervous system. And so, whatever other influences may be present, and putting aside what to Bernard was the still undiscovered role of hormones, in complex animals the autonomic nervous system plays the major role in setting the levels of life's central internal adaptive activities.

How Bernard made this discovery is a beautiful illustration of his immense talent. It had been known for some time that if one cervical sympathetic chain (there are two of these nerve bundles in the neck, one on each side) was severed, the pupil of the eye on that side of the body was constricted, while that on the other remained unchanged. When Bernard performed this experiment, he noticed something that had apparently escaped others. The animal's skin on the side of the face to which the nerve had been severed was red and quite warm to the touch.

To him this was not merely an oddity, but an important fact. He realized that the redness was due to an increase in the flow of blood to the skin of the face, and that the skin was hot because more heat was being released from the increased amount of blood passing through the tissue. But more than that, he recognized that this was the way that the body lost heat—blood flow to the cutaneous (skin) circulation.

Critically, he was able to show *why* there was more blood in the skin of the face after the nerve had been sectioned. When the nerve was cut, muscle cells that surround certain small arteries (arterioles) became flaccid. This eliminated the fall in blood flow and pressure that normally occurs when they are tense and act as clamps to reduce the diameter of these vessels. Their flaccidity had led to the swelling (dilation) of the local blood vessels and to an increase in the flow of blood through them to the capillary bed and consequently to heat loss.

But this is just the beginning of Bernard's generalizations. He realized that the sympathetic nervous system functions to constrain blood flow in *all* tissues and organs, not just the skin. The greater the contraction of the muscles surrounding the small arteries (called "sympathetic tone"), the less blood flow there is to a particular capillary bed; on the other hand, the less the sympathetic tone, the greater the blood flow. Later he showed that in certain places the parasympathetic branch had the opposite effect. It produced dilation, opening up blood vessels by inhibiting contraction of vessel-surrounding muscles.

Simply by modulating the tension of these muscles, blood flow to a tissue could be varied. From this, Bernard concluded that blood flow was not uniform throughout the body, but differed depending upon the sympathetic tone to the capillary bed of each tissue and organ. He understood that such fluctuations were adaptations designed to meet the tissue's need for blood at a particular time, and for a particular circumstance. For example, he showed that subsequent to eating a meal, blood flow to gastrointestinal organs increased greatly to accommodate the need for the digestion and absorption of food.

Finally, Bernard understood that what he was observing in regard to blood flow was an illustration of a far more general phenomenon. It was just an example, albeit an important example, of the extremely general role that the autonomic nervous system plays in regulating the body's central internal adaptive activities, from the beating of the heart to the rate of breathing. All this from a red cheek!

## *Le Milieu Intérieur*

Though it may be hard to believe, this is not the end of Bernard's contributions, not by a long shot. As it happens, the discoveries of metabolism and the autonomic nervous system have to take a backseat to his most important contribution. Through analysis and experiment he demonstrated what he called *the constancy of the internal environment,* or the *"milieu intérieur."* This was a fundamental understanding, almost on par with cell theory, about the nature of life in mammals and in complex animal and plant life more generally.

His insight encompassed nine related discoveries, understandings, and inferences, some of which are so obvious today that we have to remind ourselves that they were not in Bernard's time:

▸ The first, and the predicate for the rest, is simply that there is such a thing as an internal environment—an environment *within* complex life-forms, in addition to the one outside. As we shall see, appreciating its presence changed the way we understand life. To begin with, it made thinking of life in complex metazoan (multicellular) species in terms of two compartments or divisions—organism (cells) and environment—a serious error. There are three: cells, internal environment, and external environment.

▸ Second, the internal environment is interposed between the cells of the organism and the external environment.

▸ As such, number three: the internal environment bathes our cells.

▸ An extension of this is, four, that the immediate environment of our cells is internal, not external.

▸ And five, our cells are connected to the external environment via this internal medium. This is a paradox. The skin and the cells that allow us to perceive the external world, that allow us to hear, see, smell, and taste, have direct contact with the external world without the intermediary of an internal medium. But these are exceptions, not paradoxes. The paradox is that, in a seeming contradiction, and with only minor exceptions, the internal environment lacks *direct* access to the external environment. Interposed between the two are layers of special "epithelial" cells that face both environments and separate one from the other. In animals, they include the skin, but also the cells that line the gastrointestinal system and the lungs and gills. In this sense, there are really four compartments, not three—from the inside out, the cells of the body's internal tissues and organs, the internal environment, epithelial cell layers, and finally the external environment.

▶ The sixth understanding is that in vertebrates the internal environment is comprised of the noncellular contents of blood (the plasma) and the interstitial space (the fluid-filled space between the bloodstream and the cells of our various tissues and organs), or equivalent spaces in nonvertebrate metazoans. Together and in equilibrium with one another, they are Bernard's *milieu intérieur*, his interior environment, what we call the extracellular space.

▶ The seventh discovery was critical. Bernard's claim was not merely that there is an internal environment, but that its contents—for example, the fluid that fills the bloodstream and interstitial spaces—is of constant composition and character.

▶ By this he meant more than mere constancy, and this brings us to the eighth understanding. The physical and chemical character of this internal medium provides *optimum* conditions for the cells it bathes. It provides what they need and affords them conditions in which they thrive.

▶ This leaves Bernard's ninth, final, and for our purposes, most important discovery. He did not propose that within some statistical variation the internal environment was a static or unvarying feature of life. He understood that its contents and properties were not really constant, but rather that constancy was a goal perpetually being sought in the face of ever-present environmental circumstances that threatened to alter, disrupt, and in the end, demolish the special nature of the internal environment. The fact was that (external) environmental forces ceaselessly sought to bring the internal environment, and as a consequence, the organism and its cells, into equilibrium with it, to abolish the distinction between them.

This is just a fancy way of saying that they seek death. As Bernard referred to it, there is a "vital conflict," quite literally a matter of life and death, between environment and organism in which matter and energy are exchanged. This brings us to the central point of this discussion. It is due to the actions of life's internal adaptations that the internal environment is maintained, or more accurately, altered as little as possible, despite the resolute challenge of the external environment. They act to sustain it, and with it life, just as ceaselessly as forces external to the organism seek to dissipate it and end life. Curiously, this led Bernard to envision the conflict as being one of *benevolence*. After all, the result supported life.

And so, in complex multicellular creatures like humans, a variety of internal adaptations work to maintain the *milieu intérieur* in a particular most agreeable state, deploying diverse mechanisms toward this end and frustrating the disordering effects of external environmental forces. If they do their job well, we—both the internal environment and our cells—are to the extent possible left compositionally unchanged by environmental challenge, other than in ways that are themselves adaptive and that transpose or transform our substance and actions to be even *more* responsive to the challenges we confront.

The survival of all complex, multicellular species depends to one degree or another, in one way or another, on this effort to maintain a beneficial multifaceted internal (stationary or steady state) environment that is distinct from the environment in which they live. Sustaining it is an unending task of life. When we fail in this adaptive task, life ends.

Before moving on, we should note that I have constructed a circle, not a vicious circle, but a salutary one of mutual help and dependence, of enlightened self-interest. In it, the maintenance of the internal environment is the work of the body's cells (especially its epithelial cells, but in fact all of its cells as we shall see shortly), while at the same time and as a consequence, the medium they produce provides just the right conditions for the survival of the self-same cells.

## The Internal Adaptations of Life

As already explained, muscle, skin, and bones, as well as those aspects of the nervous system that face outward—both sensory (perception) and motor (movement)—primarily serve our external adaptations. But the rest of life's embodiment attends to our internal adaptations and are responsible for maintaining the constancy of a munificent internal environment. Indeed, for circulation, respiration, and digestion, the kidney, the autonomic nervous system, parts of the endocrine system, and even the immune system, this can be said to be their enduring purpose, their *raison d'être*. This is an enormous generalization and we should take a second to appreciate its breadth. It claims that whatever other purposes they may serve, the essential, overarching purpose of our major visceral (internal) tissues, organs, and organ systems, and the adaptations they embody, is to maintain the internal environment.

Though Claude Bernard did not talk of adaptations, he understood that to preserve the internal environment, to counteract omnipresent disintegrating, disordering, and decomposing forces, the active contrivance of the body was necessary. In order to keep alive, in their interactions with the external world, complex living things needed the means—the adaptations—to produce and sustain their internal environment.

There are two types of internal adaptations. One is responsible for carrying out the actual *exchange* of matter and energy between the external world and the internal environment. They are engaged in what Bernard referred to as the "vital work" of life. For example, the adaptive functions of the digestive tract are responsible for the uptake of solid and liquid material ingredients appropriated from the external environment (food, water, and electrolytes), those of the lungs are responsible for the exchange of gaseous matter (oxygen and carbon dioxide), those of the kidneys for providing a one-way system to rid the body of unwanted substances, those of the skin (and lungs) for heat and water loss, and finally, as its name suggests, the adaptations of the circulatory system ensure that what is taken in from the external world is made available—circulated—to all the cells of the body.

The second type of adaptation *regulates* these activities. That is, they regulate the exchange of matter and energy. By determining its rate, they assure its utility and guarantee particular, most desirable *values*—levels and amounts—for key factors in the exchange. As discussed, the central player in this regulation is Bernard's autonomic nervous system (assisted by various hormones). Together with cellular adaptations, the exchange of matter and energy between organism and environment and the regulation of this exchange comprise the internal adaptations of life, adaptations whose foremost purpose is to achieve constancy of the internal environment.

## *What Is Being Kept Constant and How?*

In the next chapter we will consider our exemplar of these internal adaptations—the means the body uses to provide sufficient quantities of oxygen to our tissues and cells—in a little more detail, but otherwise, I have not told you what is being kept constant or how constancy is achieved, other than by cells in some unspecified fashion. What are these internal adaptations and what are their underlying mechanisms? There is a good reason for my delinquency. It turns out that almost all of our contents are kept at particular values, and almost all of our internal processes—from blood flow to metabolism—are kept at particular rates. As such, merely listing them, much less explaining the underlying mechanisms that keep them constant, would be to tell you almost everything that we know about our internal workings. It would require outlining, if not reproducing textbooks in physiology, anatomy, biochemistry, molecular biology, and more, and doing so for each and every disparate group of organisms. For example, for vertebrates we would have to consider the properties and functions of all of the body's cells, tissues, organs, and organ systems—the aforementioned circulatory system, respiratory system, digestive system, kidneys, immune system, endocrine glands, parts of the nervous system (critically, the autonomic nervous system), as well as various features of skin, muscle, and bone.

Today we can list the substances that are kept constant by the hundreds, if not the thousands, each maintained at distinctive concentrations that are the consequence of its own unique circumstances. Neither Bernard nor his contemporaries knew of these molecules. In fact, his list only contained *three* items: two molecules and a physical parameter. He talked of three extrinsic requirements for animal life—humidity (water), air (oxygen), and heat (temperature)—and explained that they had to be present in the internal environment in certain quantities to support life:

▸ *Water*, the medium of life, accounts for about 75 percent of our soft tissue mass. However, it is not merely water's presence, but its state as a liquid that is critical. Liquid water is the substance in which the gases, solids, indeed all the molecules of living things, as well as the cells themselves and their anatomical structures are dissolved or suspended. Our chemical reactions occur in it and all physical movement occurs through it. And, with some interesting exceptions, life only exists at temperatures at which water is liquid, between zero and one hundred degrees centigrade. There is no solid life, no gaseous life, only liquid life.

▸ Without *oxygen*, second on Bernard's list, animals cannot extract energy from food, transform matter, or achieve (non-Brownian) movement. It is necessary to carry out the work of life in animals.

▸ And finally, three, for life's chemical processes to function effectively, the *temperature* of the body must be kept at appropriate values. Bernard was referring to homeotherms, mammals such as humans that are able to maintain a tightly regulated body temperature, but life in all organisms is critically dependent on temperature.

In the next few pages I will give three examples out of multitudes of what is kept constant. All three concern the contents of blood and each plays a critical role in the maintenance of life in complex animals. For those not inclined to details, you can skip this material without in any way losing the thread of *Life beyond Molecules and Genes*'s argument. In any event, the three things I have in mind are fundamental, indeed indispensable, aspects of living things—osmolarity, salt, and pH—and as we shall see, life depends on their constancy.

## *Osmolarity*

Though life does not exist in the absence of fluid water (spores excepted), the mere presence of water, even its abundance, is not enabling. In fact, our internal environment can be almost all water, and yet we can have too little; we can be sufficiently dehydrated so that life cannot be sustained. This is the case when an animal dies in the desert due to a lack of water. Death occurs long before the poor creature is completely dehydrated. It is not the amount of water *per se* that is critical, but the amount relative to the substances ("particles") that are dissolved in it—its osmolarity. Even small deviations in this ratio can be inimical to life. Osmolarity is usually expressed as the number of particles dissolved in a volume of water (the number of water molecules). In a given volume of water, the greater their number, the lower the concentration of water, and the higher the osmolarity. Osmolarity is "a measure of osmosis," the tendency of water to move from a region of high water concentration to one of low water concentration, or stated the other way round, from a region of low particle (chemical substances in the medium other than water) concentration to one of high particle concentration.

About 90 percent of the body's osmolarity, of the dissolved particles in cells and extracellular spaces, is due to inorganic salts, not organic substances such as proteins, sugars, amino acids, and the like, which only represent a small percentage of the total. Two types of salt predominate: sodium salts, especially sodium chloride (NaCl) ("salt"), and potassium salts, most importantly, potassium salts of chloride, bicarbonate, and phosphate, as well as potassium bound to protein.

Moreover, the osmolarity of cells and the extracellular space is the same because water moves freely between the two across intervening capillary and cell membranes. Consequently, any change in the osmolarity of blood immediately yields the osmosis of water into or out of interstitial spaces and cells. Depending on its direction, when this happens, cells either swell or shrink to maintain their osmolarity the same as blood. And this is the issue.

It is primarily to prevent such fluctuations that the osmolarity of blood is kept constant. But why should this matter? Why is it important to prevent the volume of cells from varying? As it turns out, it matters a great deal, and we can understand why with a cooking analogy. A chef's recipe can usually tolerate small variations in the amounts of the ingredients. However, if we want to increase the amount of a dish, we cannot simply add more water. We would end up with a terrible, runny mess. Likewise, to work properly, the volume of cells—the cauldrons of life—can only abide small fluctuations.

One of the first things a medical student learns is not to inject water by itself into a patient's vein. Doing so lowers the osmolarity of blood, and this leads to the osmosis of water into red blood cells, causing them to swell. If enough water is injected, they swell so much that their contents leak out across the enclosing membrane. Critically, the oxygen-carrying, red protein hemoglobin (see chapter 7) is lost, and the cells are rendered mere ghosts of their former selves, unable to function.

This is called hemolysis and the patient will die if enough of his or her red blood cells are hemolyzed. If instead our physician were to inject a *hyperosmotic* (more particles than normal per volume of water) solution, we would have the opposite difficulty.

Water would exit red blood cells and they would shrink (crenation), again leaving them unable to function properly. The concentration of their contents would rise, so that they would have too much "flour," and not enough water. And so, the medical student learns, hopefully from his or her studies and not from treating actual patients, the basic rule of medicine that any and all solutions that are injected or infused into the bloodstream must have the same osmolarity as its natural contents. Substantial fluctuations in this number are not compatible with life. For the internal medium to be hospitable to the cells of the body, its osmolarity must be maintained at a particular, essentially constant value. If the student does not learn this lesson, he or she becomes the embodiment of a dangerous environmental circumstance for the patient.

## Salt Gradients

The osmolarity of plasma and interstitial fluid is in great part (~90 percent) due to salt, NaCl, whereas that of cells is due mostly to potassium salts. This difference is critical, as we shall see. In any event, as the major source of osmolarity in blood and extracellular fluid more generally, the concentration of NaCl must be kept constant, as must the concentration of potassium salts in cells. This need for constancy also applies to salts that only represent a small fraction of the total osmolarity. For example, the potassium content of blood only accounts for about 3 percent of its osmolarity, and yet significant fluctuations in its concentration are incompatible with life.

The reason for this is that it is not merely concentration that is critical, but the *difference* in concentration (the concentration gradient) between compartments, between cells and the extracellular space, between cells and blood. Indeed, a high potassium concentration in cells *and a low one in blood* is essential to sustain life. If you enjoy reading mysteries or watching mystery programs on TV, you may have come across a story in which the perfect murder is committed by injecting a small amount of potassium chloride (KCl) into the bloodstream of an unsuspecting victim. Though an alert detective invariably figures it out, the crime is "perfect" because when the pathologist examines the blood of the deceased person, nothing abnormal is found. This is not only because potassium and chloride are normal constituents of blood, but also because the amount that is required for killing is so small that their concentration is not significantly changed, making the cause of death invisible.

Potassium is the culprit. But how does it do its dirty work? As said, it is present at higher concentrations in cells than blood—on average about an order of magnitude higher—and our bodily activities are critically dependent on this difference, on the *gradient* between them. Consequently, if the potassium concentration in blood is significantly elevated even for a brief period of time, reducing the gradient between cells and blood, cellular function is imperiled, and with it, life. Within seconds of its intravenous injection by the murderer (an unfortunate adventitious environmental circumstance if there ever was one!) or the executioner (KCl is often part of the cocktail used for intravenous execution), the bolus of potassium chloride enters the right chambers of the heart, where it elevates the potassium concentration, reducing the potassium gradient between blood and heart muscle cells. This lessens the electrical charge across these cells, and makes them briefly unresponsive to stimuli that normally signal them to contract. And of course, if heart muscle does not contract, the heart cannot pump blood, our vital tissues and organs will be deprived of oxygen, and death will ensue in short order.

To take a contrary example, how do freshwater fish, those that live in our lakes and rivers, retain NaCl at the relatively high concentrations necessary to maintain the osmolarity of their blood and interstitial fluids when the water in which they live contains almost no salt and is of low osmolarity (this is what we mean by "fresh" water)? Wouldn't the salt diffuse out of their bodies into the low salt pond? Or if that could somehow be avoided, wouldn't the salt in the fish draw pond water into the body's fluids by osmosis, diluting them, and producing fish literally bursting with water?

But neither happens and this is not because the laws of diffusion (salt out of the animal) and osmosis (diffusion of water in) are suspended for freshwater fish. Rather, they have adaptive mechanisms that allow them to regain the salt they unavoidably lose and bail out the water they necessarily imbibe. Indeed, it is *despite* these events, despite diffusion and osmosis, that they are able to maintain their salt and water content at the relatively fixed levels required to support life. Adaptations for doing this, for taking in and bailing out water, taking in and excreting salts, are central life activities for most animals. Indeed, it has been estimated that vertebrates expend between one-third to half their energy merely to maintain this balance.

## *pH*

Finally, let us consider a factor related to chemical constancy—pH. pH is how basic or acidic something is. At a pH of 7.0, a medium contains equal amounts of acid and base, $10^{-7}$ M (moles per liter) of each. A pH of less than 7 is acidic while a pH of more than 7 is basic or alkaline. pH is a logarithmic parameter so that a pH of 6 is ten times more acidic, and a pH of 8 is ten times more basic than one of 7. Human blood is slightly basic, pH 7.4. As a consequence, the concentration of acid in our blood is a little less than $10^{-7}$ M (0.1 micromoles per liter [pH 7.0]), or about 0.06 micromoles per liter. Like blood, the pH of the cell is roughly neutral (somewhere between pH 6–7). The pH of a relatively dilute solution (0.1 M) of a base like sodium hydroxide (NaOH) is 10, about three orders of magnitude more basic. As for acidity, to give you a sense of how little acid this is, purified water contains an order of magnitude more, orange juice about four orders of magnitude more, and an active stomach busy digesting food, six orders of magnitude, or a million times more.

In any event, given the relative neutrality of our bodily fluids, it should not come as a surprise that most biological reactions occur, and are maximally effective, at neutral pH. Indeed, life cannot tolerate even small variations in pH. Blood pH normally varies in the hundredth place, for example from 7.38 to 7.42. If the acidity of blood is increased (or decreased) by as little as 0.02 micromóles/liter, we cannot survive. And yet when cells do work and use energy, they produce more and more acid. This is because the end products of oxidative—that is, oxygen-requiring metabolism—are carbon dioxide ($CO_2$) and water ($H_2O$), which when combined produce an acid, carbonic acid ($H_2CO_3$).

Indeed, the body produces prodigious amounts of acid, and consequently we would expect the pH of blood to quickly fall outside the very narrow limits that can sustain life. But this does not happen. The pH of the body remains doggedly neutral. The body is able to rid itself of the acid it produces, mole for mole. It has the means, the adaptive mechanisms, to excrete what it makes. First we get rid of gaseous $CO_2$ by exhaling it, preventing its combination with water in the body, and second we excrete what remains, or its equivalent, in our urine.

## Homeostasis

Early in the twentieth century, Walter Cannon, a professor of physiology at Harvard University, expanded on many of Claude Bernard's ideas in his research and in a wonderful and important book titled *The Wisdom of the Body* (New York: W. W. Norton, 1932). In it he provided what for its time was a rich account of life's internal adaptations, of what is kept constant and how constancy is achieved.

Cannon coined the term "homeostasis" to describe this phenomenon—the equipoise of the body, the tendency of all of its internal activities toward some normative sustained value. The word "homeostasis" comes from the Greek roots *homoios* meaning "like" and *stasis*, meaning "standing"—standing together in like form. He not only envisioned homeostasis of the *contents* of the internal environment, of blood and interstitial fluid, but of all of the body's internal activities. Each he said has a normal value or physiological rate, and as such is maintained in some sort of "equilibrium," not meaning physical equilibrium, but balance.

Such values or rates indicated proper function, or normal physiology. When we get the results of clinical tests on our blood or various bodily functions, we hope that our numbers will be in the "normal range," the same as that for all well-functioning individuals of our species within some predictable statistical variation. This means that at least in regard to the things being measured, our cells, tissues, organs, and systems are functioning properly both in their own right and in their efforts to maintain a stable internal environment. Other symptoms notwithstanding, if a series of important values are in the normal range, we are certified healthy. If they are outside that range, then, to one degree or another, we are in poor health; and our homeostatic, our adaptive mechanisms have been compromised.

Despite the common wisdom (human, not that of the body) that we are what we eat, no matter how much or how little fat I take in, my body does its best to ensure that the lipids in my blood are kept at a "normal" level, not more, not less. Even if I eliminate most of the salt from my diet, the salt content of my blood does not change. My body adjusts; it adapts to its circumstances. It maintains homeostasis. Today we know that this applies to countless substances and processes that Claude Bernard could not have imagined.

As Cannon explained, the body seeks this most salubrious state for every significant internal feature of life. Almost everything about us is maintained at desirable or advantageous values by adaptive mechanisms. As said, this not only applies to our contents—to osmolarity, pH, salt, fat, and the like—but to how much blood the heart pumps, at what pressure and frequency, the rate of blood flow through our various blood vessels, its distribution among our diverse tissues and organs, as well as how much air we breathe and the composition of what we expire—in sum, every process that affects the constancy of the internal environment. As such, a multitude of adaptive processes designed to maintain one aspect of this constancy or another—each homeostatic in its own right, each occurring at a most beneficial rate—are hallmarks of the living state.

Finally, Cannon stressed that constancy without the possibility of change does not produce homeostasis. Homeostasis is not merely the body's attempt to achieve certain constant values. These values must be capable of varying in a purposeful manner in response to new needs, new external circumstances. That is, they may be paradoxically elevated or decreased, changed, not kept constant, in order to maintain homeostasis. For example, during a lion's attack on its prey, the distribution of blood is very different than afterward. During the kill, blood is rushed to the muscles, and away from the gastrointestinal system. Afterward, as the prey is being devoured, it is directed to the gastrointestinal tract and away from muscle.

To produce such changes as well as to maintain particular stable values and rates, *regulation* is required; that is, the body must act in one way or another to change or preserve particular values in the face of an environmentally provoked challenge. As mentioned, Claude Bernard was the first to appreciate that physical parameters such as blood flow are regulated, and it was none other than Ivan Pavlov who was the first to realize the necessity of regulating the rates of chemical reactions. But Cannon put it together. He not only described the constancy of the contents (and temperature) of the internal environment, but the regulatory adaptations—those of the autonomic nervous system, the nervous system otherwise, and hormones—designed to assure it. As he explained, the wisdom of the body, homeostasis, does not reside in some existential constancy, but in the means, the adaptations that produce and maintain it.

## Darwinian (External) versus Bernardian (Internal) Adaptations

As I have here, Cannon divided life into two realms: internal and external. The nervous system, he said, has two major branches, one that faces outward, "exterofective," and one that faces inward, "interofective," each containing sensory and motor arms. But he also understood with Bernard that more generally there are two types of adaptations—internal and external. Though both deal with the challenges presented by the environment, they do so in different though complementary and, indeed, often linked ways.

Adaptive functions in the external world affect those in our internal world, and vice versa. In the example of the lion attacking its prey, though the lion uses Darwinian adaptations to obtain and ingest food, doing so activates a variety of internal adaptive processes needed to process the meal and make use of its contents, from the secretion of digestive enzymes to increasing motility along the gut to a changed distribution of blood. Not only that, but the lion was motivated to seek the food in the first place by an internal adaptation—hunger.

Similarly, when an animal is hot, the blood vessels in its skin dilate to help it lose heat and prevent its body temperature from rising. In some vertebrates, like dogs, heat is also lost through the respiratory tract by panting. On the other hand, if it is cold outside, the blood vessels in the skin constrict to better retain heat (and panting behavior ceases). These are internal adaptations that enable us to keep body temperature constant despite variations in environmental temperature. But we can also react to changes in environmental temperature by engaging *external* adaptations. For example, if we are hot, we can take off clothing, avoid the sun, or change our posture by spreading out to expose more surface area, or if we are cold, we might do the opposite and put on clothing, seek the sun, or change our posture by curling up.

For the most part, external adaptations deal with the inconstant exigencies of our environmental circumstances, whereas internal adaptations are responsive to their abiding if sometimes varying physical and chemical character. The former adaptations act directly to ensure survival by, for example, seeking food, while the latter adaptations work indirectly by maintaining our culture medium, our broth, our nurturing internal environment in a most desirable state.

And so, internal adaptations seek to produce a tranquil and beneficial interior world that is independent of the vagaries of the external environment. As such, along with what Bernard referred to as "the nutritive reserve," such as the storage of glycogen and fat, they not only help us overcome immediate challenges presented by our external environment, they enable us to leave adverse and unwelcome situations and unfavorable environmental niches in search of more harmonious circumstances. It is said that we carry our "environment," our internal environment, with us.

CHAPTER 7

# The Physical World Intrudes

## *How the Physical World Shapes Our Adaptive Properties and Us*

---

[Life] cannot be explained by an internal principle or action; it is the result of a conflict between the organism and the ambient physicochemical conditions.

—CLAUDE BERNARD, *LECTURES ON THE PHENOMENA OF LIFE COMMON TO ANIMALS AND PLANTS*, 1878

THERE IS another extremely important difference between external and internal adaptations. According to Darwin, in addition to reproduction, evolution depends critically on two factors: variation and natural selection. Variations are differences between organisms—differences in their incarnations, in their character or properties. It is from the action of natural selection on these differences that life's diverse adaptations evolved. Both variations and natural selection are commonly understood to be matters of chance. Variations arise from *random* genetic mutations, and natural selection results from *chance environmental circumstances and encounters*. As a consequence, evolution's path is accidental, a matter of chance; it is not predetermined.

And yet, as we learned in chapter 5, for at least one situation—the emergence of the circulatory system—it was not a matter of chance, the result of random events, but rather it was *destined and inevitable*, at least if organisms as large as humans were ever to exist. Its evolution was the unavoidable consequence of fundamental physical and mathematical laws and rules—in particular, those of diffusion and geometry.

## Getting There, So As to Get There

The circulation is not alone in this, not by a long shot. Unlike Darwin's external world—the world of adventitious circumstance—the *rules* of the environment—nature's laws, not chance—are responsible for many, if not most, of life's internal features. They are responsible for Bernard's internal world. As explained, to describe this world for even a small sample of its features would require a massive book, the size of many textbooks combined. Fortunately, one good example is sufficient to show how deeply embedded nature's laws are in the outcomes of evolution.

Consider how large animals get the oxygen they so urgently need, and what is required to achieve this goal *beyond vascular perfusion*. For vascular perfusion to be of any use, the blood must contain the desired oxygen, and for this the gas must be extracted from the air and placed in blood. That is, it must first get into blood to be perfused through our various tissues and organs.

But how does this occur? How does oxygen get into blood? The central constraint is the distance between the two, between air and blood. For the exchange to be effective, the distance between them must be very short—indeed, microscopically short, about ten to twenty microns (0.0005 of an inch), the thickness of one or two cells. This is the law of diffusion at work. If the distance were much greater, oxygen could not enter blood rapidly enough to meet our needs. Remember that diffusion becomes an increasingly inefficient process as the distance traveled lengthens, even across so thin a barrier as a strand of hair or the thickness of several cells.

Because of this unalterable fact of our environment, blood vessels and air must be cemented to each other like two lovers separated only by their respective skins. And the material that is interposed between them is in fact the skin in which the blood vessels lie. But even a grade school student knows that oxygen does not get into blood across our skin. We do not breathe through our skin. We breathe through our lungs.

But why don't we breathe through our skin? Certainly it is the most direct way of exchanging gas between the body and the atmosphere. Why doesn't oxygen simply diffuse into the suffusion of thin-walled blood capillaries located just beneath its surface? Why didn't evolution choose this simple and direct means of obtaining oxygen in us as it did in many other species? Or more accurately, why didn't nature *stick* to this means of obtaining oxygen? Why go to all the trouble, and trouble it was, of making lungs (or gills for fish) and the complex system ancillary to and required by them?

As with the circulation, nature had its reasons and they were not merely compelling, they were inescapable. For diffusive exchange to occur across the skin, the skin would have to be, as said, about 0.0005 inches thick, or microscopically thin. A person with such a thin skin would be defenseless, harmed by contact with anything but the smoothest surfaces, constantly being cut and abraded, and extremely susceptible to deadly hemorrhage and infection.

As the skin actually evolved in most vertebrates and in many invertebrates, it was *not* to facilitate the transfer of gas from air to the body, but rather to serve as a *protective barrier*. Though its structure varies greatly from species to species, the external surface of an animal, its "skin," is usually comprised of one or more of the following: numerous layers of cells called the stratified (many layers) squamous (flattened cells) epithelium, overlaying layers of dead skin cells, inanimate material such as mucous and keratin, or in higher invertebrates an exoskeleton of shell, carapace, and the like, all there to, among other things, protect the body from harm.

But even if organisms with a very thin skin could somehow survive life's vicissitudes (small invertebrates do), this would not solve the problem. There is another insurmountable difficulty.

The amount of oxygen that can dissolve in the water of the body's surface layer, the preface to diffusing into the underlying capillary bed, is critically dependent upon the area available for this exchange to take place. The greater the surface area is, the more oxygen that can be exchanged. Though larger objects obviously have a larger surface area than smaller ones, the rules of geometry tell us that the *ratio* of surface area to the underlying volume or mass of the object becomes increasingly *small* as size increases, making things less, not more, proportional.

Although the geometry of living things is complex, irregular, and variable, nonetheless the relationship between surface area and volume is much like that of a sphere. The surface *area* of a sphere increases with size as the square of its radius, while its *volume* increases to the third power. This means that as size (volume or mass) increases, surface area increases to a lesser and lesser degree. This simple rule of geometry, an environmental attribute, has played a critical role in determining the form of life on all scales.

And so the *area* available for gas exchange via the organism's surface *decreases* in relative terms in an exponential fashion as the object becomes larger. Yet due to their increased mass, larger animals have a greater need for oxygen than small ones. In fact, animals far smaller than us have too little skin, too little surface area, to exchange gas effectively given their size. A sufficient surface to volume ratio is only found in microscopic organisms, in single cells or thin layers of cells. As it happened, as the size of organisms increased and the need for oxygen became *greater*, evolution was presented with a limiting problem.

## The Lungs

If large animals were to evolve, there had to be some way to exchange gas other than across the skin. But how could this be done? What, like a thin skin, could provide the necessary proximity between air and blood, but still protect the animal from the danger of cuts and abrasions, and in addition provide a sufficient surface area for effective gas exchange? Nature's solution was brilliant, though it would be more accurate to call it fated.

A piece of skin was internalized, brought *inside* the body as an in-pocketing of the surface (actually, an in-pocketing of the pharynx, a piece of skin that had already been brought into the interior of the body for the purpose of ingesting food). Being inside the body, this piece of skin was protected from mechanical trespass from what lay outside, protected by the outer skin and interposed tissues and organs. As a consequence, the gas-exchanging surface could be very thin without constantly being traumatized.

Moreover, because the internalized "skin" was within the body, it could not spread out (or proliferate) in a planar fashion as it would on the surface, enclosing larger and larger volumes as it did, therefore making the situation worse, not better (by reducing the ratio of surface area to volume). Tissue growth *inside* the body, being constrained by the enclosed space, turned out to be a critical benefit, not a disadvantage.

By physical necessity, the growing tissue had to turn back on itself again and again, scrunching up into a ball-like structure that conformed to the shape of the confining space (our chest cavity). As a consequence, the surface area of the internalized skin could be greatly increased without producing an even greater increase in the volume it enclosed. To visualize how this occurs, imagine six pieces of ordinary stationary paper taped together loosely to enclose a cubic rectangle. It is easy to greatly reduce the volume of the constructed object. Just crumple the paper up into a small ball and express the enclosed air. But of course this does *not* reduce the surface area. It is still the sum of the areas of the separate, now crumpled, sheets of paper.

By favoring the internalization of a piece of skin, natural selection had achieved three crucial objectives: (1) it allowed organisms to protect their gas exchanging surface, thereby (2) permitting it to be extremely thin, while at the same time (3) accommodating a great increase in surface area that did not simultaneously increase the enclosed volume even more. All of this simply to conform to environmental demands, to the laws of diffusion and the rules of geometry!

There was one further requirement, and it was as critical as it was obvious. For the internalized piece of skin to be of any use for gas exchange, direct contact between it and the surrounding air had to be maintained despite its location inside the body. However crumpled, the cells of the internal surface had to have direct access to the air outside. For this to occur, the *environment* had to be brought *inside* the body. In mammals and in other vertebrate lung breathers, as well as in many invertebrates, a series of tubes or ducts—the trachea, bronchi, and bronchioles—produced from certain cells of the internalized layer, accomplish this task.

In lung breathers, these tubes lead to and end in multitudinous closed-off sacs called alveoli that are comprised of a single layer of very thin cells that are closely apposed to a vast underlying lake of capillary blood. To ensure efficient exchange, at the same time as the lungs formed, blood circulating through the lungs, which are only a small fraction of the body's mass, evolved to equal that circulated through the rest of the body all told. That is, unlike all other organs, blood circulation through the lungs is in *series* with, not parallel to, that through the rest of the body.

## Breathing

But as often happens, in solving one problem, another is introduced. The ducts that bring air to the alveoli are long and tortuous, and *diffusion* through them from the air into the deepest recesses of the internalized layer, the alveoli, would be ineffective. It would take hours to carry oxygen to the exchanging surface. And so, it was not enough just to internalize a thin piece of skin and form a structure with a large surface to volume ratio. A mechanism was needed to carry air through the long bronchial tubes to the layer of gas-exchanging cells.

Once again, the solution was mandated by the properties of

the physical world. Remember that nature only provides two ways for molecules to move—diffusion or flow. Since diffusion couldn't do it, some way was needed to produce the *flow* of air through the tubes to connect the environment to the alveoli. Just as with blood flow, this required a pressure gradient—this time of gas, not fluid.

In the circulation of blood, the task of producing the gradient was given to muscle contraction (the heart). Likewise, for breathing, the muscles of the rib cage and diaphragm act to move air into and out of our lungs. When we breathe in, these muscles expand our chest cavity up and outward, increasing the volume the lungs can occupy. This in turn reduces the pressure of the gas contained in the lungs below that found in the atmosphere (the same amount of air is now distributed in a larger volume). As a result, gas (air) flows from the higher pressure atmosphere into the lungs, equalizing the pressure between the two as it does. When the inspiratory muscles relax (or when we engage our expiratory muscles), we exhale. The chest cavity now moves down and inward, and the volume of the lungs is reduced. This increases the pressure in the lungs (the same amount of gas in a smaller volume) so that the direction of gas flow is now reversed and air in the alveoli is expelled into the lower pressure atmosphere.

And so, a complex system of muscles and supporting bones (the rib cage) evolved to produce the flow of air required to ensure gas exchange at the alveoli, all mandated by physical and geometric laws. But as with locomotion, this was not enough. Something was needed to make sure that the muscles contracted at a rate and that breathing occurred at a depth needed to guarantee adequate amounts of oxygen in blood (as well as to ensure the exhalation of sufficient quantities of $CO_2$). This responsibility fell on the autonomic nervous system. It innervates the muscles of respiration as well as various receptors strategically located throughout the body—at portals to the brain and the heart itself—that monitor the effectiveness of breathing by measuring the oxygen and $CO_2$ content of blood.

## The Impact of the Physical World on the Outcomes of Evolution

But this is hardly the end of the story. For example, merely dissolving oxygen in blood, like salt in water, cannot do the job. Not enough gas can be dissolved at the temperature and pressure of the earth (and of blood) to provide our cells and tissues with the amount they need. Indeed, even if everything I have described to this point was available and functioned perfectly, it would all be for naught, unless there was some way to increase the amount of oxygen blood could carry.

Humans and our vertebrate relatives have a special protein—hemoglobin—that natural selection has designed to overcome this difficulty. How this occurs chemically and physiologically is a complex business that you can learn about in any good textbook of biochemistry or physiology. Suffice it to say here that hemoglobin binds oxygen avidly and as a consequence guarantees sufficient amounts in blood to meet the needs of our cells and tissues.

I could continue and provide in greater and greater detail a description of the physical and mathematical constraints that determine the intimate workings of the complex and elegant systems that provide oxygen to our cells and tissues. For example, it is not enough for hemoglobin to just grab oxygen—it must relinquish it at its cellular destination, and a mechanism is needed for this. But no matter how detailed or comprehensive I might get, almost everything about the means the body uses to provide its cells and tissues with oxygen has been predetermined by the demands of the physical world in which we live, that is to say, by our environment.

This fact of life not only applies to vascular perfusion and respiration, but to most, if not all, of life's internal activities. Consequently, the way we are is not merely a matter of chance. In more ways than we can count, the path of evolution was constrained and determined by the predictable physical properties of the environment.

To say this is not to ignore the role of chance in evolution. The evolution of our *external*, Darwinian adaptations—those that face outward—has in great part been predicated on unpredictable or only statistically predictable qualities of our environmental circumstances, such as weather, attack by predators, and all the other capricious occurrences of life. They have certainly been central to the evolution of the species, and indeed, if we counted up all of our adaptations, internal and external, perhaps most would be derived from chance events and function in their service.

When we think of such causes, natural selection seems fickle and erratic. But for most of the internal characteristics of the organism—its inwardly facing adaptations—natural selection is not unpredictable at all. Rather, it reflects and is dependent upon *constant* features of the environment, such as the laws of diffusion and rules of geometry. In these instances, the outcomes of natural selection, though not always easy to predict, are entirely predictable. This is also the case for some outward-facing adaptations—for example, those that produce movement. From muscle to bone to the means of coordinating movement, it is all the consequence of nature's enduring laws of mechanics.

In any event, almost everything about the means the body uses to provide adequate amounts of oxygen to its cells and tissues—getting oxygen into blood (extracting it from air), accumulating it in blood (bound to hemoglobin), presenting it to the body's tissues and organs (vascular perfusion), and unloading it at the tissues (release from hemoglobin)—is dictated by the physical nature of the world in which we live, by its circumstances, by the laws of the environment.

I could tell parallel stories about the gastrointestinal tract, the kidneys, and other tissues, organs, and systems. Why is the GI system a long tube with a musculature to move material through it? Why is the kidney comprised of millions of minute tubes lined up in parallel? The answers to such questions can ultimately be traced to the physical requirements of the world in which we live.

The modern focus on molecules, especially on DNA and proteins, makes it seem otherwise. After all, *their* structures and the functions they embody are the result of chance variations, random mutations that just happened to have had advantageous functional consequences. For DNA these advantages concern the code, and for proteins their reactive and binding properties. From this point of view, together they give rise to life's characteristics, making natural selection no more than a fortuitous event among randomly produced variants.

However, for the major internal adaptations of life, this is far from the case. Although variants may be randomly produced, the bases for natural selection's choices are not random. If you think about it for a minute, you will see that this is not even true for the protein molecule itself. Though its structure may be the happy consequence of random events, what it *does*, more often than not, is not a matter of chance, but physical and chemical necessity. And as we shall see, it is what it *does*, not what it *is*, that natural selection acts upon and upon which the path of evolution is predicated.

This said, and the physical laws of nature notwithstanding, animals have traveled down different evolutionary paths. They have evolved alternative means to achieve the same ends. Do we breathe in air or water? Do we walk, fly, or swim? And on and on, feature by feature. But it is important to keep in mind that whatever path has been chosen, and however the processes and mechanisms of life differ, they were shaped in one way or another, to one degree or another, by the physical characteristics of the world in which we live, by predictable, universal, and omnipresent circumstances, not just periodic chance occurrences.

It was to accommodate these environmental conditions that life's internal adaptations evolved, allowing and sustaining life in all its variety. Unlike most Darwinian adaptations, these Bernardian adaptations were not the result of the uncertainties of our circumstances, but of its constancies. But this does not make our heart, blood vessels, or lungs any less of an adaptation than the shape of beaks.

# CHAPTER 8

# Adaptations as Life

## *Adaptations as the Sufficient, Lifegiving Properties*

---

A biologist has no right to close his eyes to the fact that the precarious balance
between a living being and its environment must be preserved by some
mechanism or mechanisms if life is to endure. No coherent attempts
to account for the origin of adaptations other than the theory of
natural selection and the theory of the inheritance of acquired
characteristics have ever been proposed.

**—THEODOSIUS DOBZHANSKY, *GENETICS AND THE ORIGIN***
**OF THE SPECIES, 1937**

IN THEIR SEARCH for life's causes, neither Aristotle nor Galen
*inter alios* sought to explain differences between the living and
the inanimate as being of a material kind, like those between
apples and oranges. It was not that they did not think that such
differences existed or even that they were important, but rather
they were confident that they were not sufficient to explain life.
Life was a *state*, not merely a material embodiment, and this
was self-evident not merely to deep thinkers like Aristotle and
Galen but, as already explained, to anyone who had seen death.
Since an organism before and after its death is the same object,
it seemed self-evident that death involved a change of state. The
question was what was that state? Since death was the *loss* of life,
it seemed to involve the *loss* of something: something that animated the object, that gave it life.

The word "animate" comes from the Latin word *anima*, breath or soul, and means "gives life." Breath/soul was the lifegiving attribute. Breath was or contained soul. When we stop breathing, we are no longer being infused with the life force or substance, and consequently we die. It did not matter whether one thought of *anima* as material or immaterial, substantial or insubstantial, whatever it was introduced the living state, whatever that was.

Though modern science has excoriated the notion of soul or *anima*, completely extinguishing it from the scientific lexicon, it has yet to be able to replace it. And though many scientists today believe that terms like "genetics," or "DNA," or "chemistry" are proper scientific replacements, if you have read this far, you should realize that this is not the case. Even the wondrous DNA does not do the job. And so, the question remains: what change in state occurs when we die? What exactly is lost?

## Just Being Alive

We have learned that my apparent disengagement from the struggle for survival as I relaxed in my reclining chair was an enormous misperception. Even if I were somnolent, much less asleep, and contact with the external world seemingly all but cut off, within my body multitudes of adaptations would have been hard at work resisting various life-threatening challenges presented by the environment and keeping me alive.

There is another closely related misperception here, and it is as powerful as it is empowering. As I sat in my chair, I felt alive inherently, simply by being. It seemed that the things that made me alive and the adaptations involved in the struggle for survival were completely different phenomena—the first imparted life, whereas the second were features of otherwise living objects. It was my unmistakable impression, my unequivocal perception that I stood (or sat) on my own as a living being distinct from the world in which I live and with which I interact. As I sat in my chair, my aliveness seemed intrinsic.

And yet, as should be clear by now, the fact of the matter is that my state of aliveness not only depends on the effective deployment of adaptations contrived by evolution to defend me against environmental challenge, it completely depends upon them. In their absence, in the absence of adaptive responses to environmental threat, there is no life, no aliveness. And, as explained, death is nothing but the loss of these life-endowing adaptive properties.

Critically, there is no physical or chemical feature or group of features that bestows life on an object independent of this essential interaction. Not only that, but however counterintuitive it may seem, my adaptive properties do not exist because I am alive, I am alive because they exist. They are life's true causes. As said at the outset, this is the central claim of this book. Biological adaptations are soul, *anima*.

In addition, it is claimed that adaptations are not embodied in, not made flesh by, a particular material incarnation. They are contingent phenomena, and as such only exist in action and in context. When the cast of an organism matches its surroundings, hiding a vulnerable species from predators, its color is adaptive. But when its color does not blend with a particular background, no adaptation exists. Without bees or other insects to pollinate the flower, its color provides no adaptive advantage.

And as said, for me sitting passively in my chair, not aggressing nor being aggressed against, not seeking to produce offspring, or anything else, indeed not seeming to face any external threat whatsoever, I appear adaptationless both contextually and in (in)action. I am not in harm's way, and my adaptations only exist when I am. Though we have learned that this is not really true because my internal adaptations are hard at work defending me against unseen threats, it is true to the extent that various potential threats to my survival are not at play. Consequently neither are the adaptive properties designed to deal with them.

If my lifegiving adaptations are contingent, not inherent phenomena, then the properties that make me alive do not exist in some sort of splendid intrinsic material isolation, and life is not an autonomous, but a dependent phenomenon. Indeed, it is wholly dependent upon these contingent interactions with the environment. And if life depends on adaptations, and adaptations are contingent phenomena, then life itself is contingent, not intrinsic.

## The Living Thing Itself

Yet how can this be correct? In the absence of most adaptations, life endures. With critical exceptions such as circulation and respiration, their absence may threaten my existence, I may be less likely, perhaps far less likely, to survive life's challenges, but I remain alive nonetheless.

For example, though its speed is a crucial adaptation to escape predators, the gazelle is not alive *because* of its speed. A lame gazelle seems no less alive than a fit one, as is a bird with a broken wing or a human paraplegic. Despite failure of their ambulatory capacities and their increased vulnerability, despite their infirmities, they remain alive. If we were to limit Galapagos finches to food sources that do not depend upon the various shapes of their beaks, they would still be alive, though their beaks would no longer serve this adaptive function. A plant without its protective thorns or odious taste would continue to live, though it might be more likely to be eaten. And certainly an animal that exhibits protective coloration when sitting on a like-colored stratum does not die if it moves to one of contrasting color.

The same applies to most internal adaptations. If my ability to regulate my blood pressure is compromised—say I have primary hypertension—I might be more vulnerable to environmental stress, but am no less alive as a consequence. Or if my wound-healing ability is impaired due to diabetes (abnormally high blood sugar levels), I would still be alive unless or until this maladaptive situation leads to a mortal infection or some other terminal event. And in spite of studies that show that abnormally high cholesterol levels in blood make me more susceptible to coronary occlusion and a heart attack, until these horrific events actually happen and lead to death, I remain alive. Whether these difficulties reflect a diminished but still present adaptive power, a total lack of adaptive utility, or are even maladaptive, I continue to be alive.

But even more important, much, perhaps most, of my external adaptive potential is not being utilized at any given time. If life is contingent on its adaptations, and adaptations are contingent phenomena and hence do not exist when not in use, then why do I not seem to be in the least diminished by their absence? I seem just as alive. And if I am fully alive without some, much less many, of my adaptations, then how can the claim that they are life's causes be true?

Yet I submit that they are. Even if I could somehow imagine a "biological" being without adaptive properties and yet able to do everything that living things do, it would not be alive. But I cannot imagine such a being, because no such creature could exist. We are alive because of the special means we have been given by evolution to *adapt* to our circumstances, to engage our environment and to persevere in an unkind world. Each and every biological adaptation, and in a sense even their mere potentiality, is an aspect of aliveness. And if their absence is not compensated for by other adaptations or by a less stressful environment, we are indeed diminished, however inconsequentially, by their absence or impotence one and all. Life is not a single state, discontinuous from all others, but a continuum of states, of varying degrees of adaptive competence, and consequently aliveness.

## Material Incarnations

If life is not intrinsic to its material incarnation, but is the consequence of adaptive properties that transcend it, even those that only exist in the world of potentialities, we should be able to separate life's material embodiment from its adaptive characteristics in some fashion. But can this be done? When you get right down to it, aren't they part and parcel of the same thing? Contrary to what I have said, doesn't our material incarnation *embody* our adaptations? Indeed, doesn't this seem self-evident? What else could embody them? In fact, isn't it true that changes in the material nature of living things from life's origin to the present day do not merely appear to be an *aspect* of evolution, but evolution itself?

Isn't evolution as much about the unfolding of life's *material* nature as its adaptive character? Today's DNA is a far more complex substance that contains incomparably more information than the primitive DNA that first came into being billions of years ago. Likewise, the number and variety, much less the complexity, of protein molecules today is far greater than it was at the beginning of protein chemistry. In a more general sense, today's biochemical reactions are more specific, part of far more complex chemical cycles and sequences, than when biochemistry first began. Nor were the structures of the modern eukaryotic cell present at life's beginning. Finally, and most obviously, the remarkable multiplicity of form of multicellular organisms with their many shapes and sizes, their different structures and anatomical arrangements, were not present at life's beginning. If we know anything at all about evolution, it is that our material embodiment evolved from what it was to what we see today!

And so, if it is also true that only adaptive features evolve (having been chosen by natural selection), then aren't life's material incarnations and its adaptations one in the same thing? If through natural selection, evolution simultaneously produced both our material being and our adaptive nature, how could our materiality have evolved without being *intrinsically* adaptive? And yet, it is not. Though closely related, deeply consanguineous, interdependent, they are not the same thing.

And it is simple to see that this is so. As already explained, the clear and unambiguous fact is that any and all aspects of my material incarnation can exist *without* expressing an adaptive function, most absolutely on death. As such, they cannot be adaptive intrinsically. A particular material embodiment may be prerequisite for the expression of a specific adaptive property, but the presence of that embodiment does not guarantee the presence of the adaptation.

This said, it is of course our material constitution that makes our adaptations to nature's challenges possible. It is their corporeal basis and, as such, is prerequisite for their existence. But this does not make life's material aspects adaptive *inherently, in and of themselves, by dint of their being.* Or putting it the other way around, our adaptive properties are no more intrinsic to our material nature than they can exist independent of it as disembodied phenomena. Though two peas in the same Darwinian pod, they are not identical, not the same thing.

## *Life Itself*

As it turns out, to understand this difference is nothing less than to understand the phenomenon of life. When scientists talk about life's "emergence," they often leave the term undefined or defined by analogy to something else. For instance, the emergence of life is often likened to the chemical and physical properties of a molecule that *emerge* from the atoms that comprise it. The atoms themselves do not display these properties, but the molecule does. The analogy certainly applies since we are comprised of molecules. But, as discussed, chemical emergence is *not* the same phenomenon as the emergence of life. *Life's emergence is the coming into being of adaptive properties in all their variety from their material underpinnings.* This is not analogous to the coming into being of chemicals from their constituent parts or the products of reactions from predisposing reactants. Underpinnings are not merely or necessarily constituent parts, and adaptive properties are not merely or necessarily chemical properties. In fact, they need not be anything like them. For example, underpinning the adaptive property of tissue and organ perfusion is the flow of blood, and blood flow is neither a constituent part nor a chemical property or reaction.

Because life's material features are not adaptations in their own right, they cannot be *sufficient* causes of life. Neither DNA, protein molecules, the other chemicals of life, nor the reactions in which they participate, their interactions with each other, not even the anatomical structures in which they are found and which they form, are biological adaptations *in and of themselves*, even though they are absolutely *necessary* for their expression. Adaptations—life's sufficient properties—emerge from its necessary ones, but, that said, unlike atoms and the chemical compounds that emerge from them, the necessary and the sufficient properties of life are not merely distinguishable from each other, not merely different, but quite remarkably are most often nothing at all alike.

## The Selfish Gene

Drawing this distinction between our materiality and our adaptive character is not some sort of philosophical splitting of hairs, but is central to understanding the phenomenon of life. The only adaptive advantage that a particular molecule of DNA offers in and of itself is *to itself*, its physical stability, just like a rock. And the reactions it can engage in offer no further advantage beyond the reactions themselves, and they are merely chemical reactions like all others.

The only thing DNA can *do* by itself, or in combination with other chemicals, is produce particular chemical products, either by being broken down in the process, being added to, or simply acting catalytically with no change in its structure. Only *within* the organism does it serve as the underlying material basis for a wealth of life's adaptive properties. And this is not just an irrelevant coincidence. It is critical to the notion of life.

Nor, though a popular notion today, did DNA's evolution from a relatively simple chemical polymer to the incredibly complex substance of the Genome Project come about because particular forms of DNA were favored by natural selection. Rather, DNA served as the ultimate material foundation for the various protein-based adaptive properties that evolution produced. Differences between the DNA molecules of different species gave rise to certain advantages as well as weaknesses in the *organism* as it struggled for survival. But the struggle was the organism's, not the DNA's. The advantages and disadvantages that became manifest were not those of DNA, they were not even those of its encoded proteins, but of the organism.

This, as Darwin explained, is because adaptability to circumstance is a feature of the whole organism and it alone. Paradoxically, this understanding was challenged as a result of the attempt to bring Darwin's theory of evolution and Mendel's theory of "genes" into harmony in what is referred to as the "modern synthesis" or "neo-Darwinism." When Mendel's observations on the pea plant were rediscovered in 1900, many thought his explanation of heredity was in fundamental conflict with Darwin's theory of evolution. It seemed that either genes or natural selection were at work, but not both.

The conflict was eventually resolved by explaining natural selection in terms of genes. It was proposed that natural selection acted on genes—understood to be the underlying basis of life's features—favoring some, discarding others. With the discovery of the genetic code in the DNA molecule, this equation took on a very specific material meaning. Genes were particular sequences of nucleotides in the DNA molecule, and it was these chemical sequences that were the substrate for natural selection. And if evolution is the evolution of adaptations, then adaptations must be properties of DNA.

In recent times, two wonderful popular science writers on the subject of evolution—Richard Dawkins and the late Stephen Gould—argued about the locus of the adaptive properties of life, resurrecting arguments that date back to the beginnings of neo-Darwinism in the 1930s. In his first book (1978), *The Selfish Gene,* and in those that followed, Dawkins argued that natural selection acts on DNA, not organisms. And that as a consequence, and as just described, the adaptive properties that resulted—the outcome of selection—were embedded in DNA and its genes. Essentially, genes in DNA are what evolution is all about. DNA evolved; our cells and bodies were mere receptacles for this substance.

Gould had a far more expansive view of adaptations and natural selection. He argued that while selection occurred at the level of DNA, it also occurred at other levels of organization. Adaptations were not only found in DNA, but in parts of cells like mitochondria and the endoplasmic reticulum, in the cell as a whole, in the whole organism, and even in populations of like organisms.

Before we draw a conclusion about these opposing points of view, we need to be clear what we mean when we say that something is an adaptation. Needless to say, it was not without thought that Darwin concluded that natural selection acts on whole organisms, not bits and pieces of them, and that adaptations were *their* properties, and their properties alone. Indeed, this is what he meant by natural selection—a selection between *organisms* based on their fitness, or their adaptive capacity.

This understanding was foundational. The only way that DNA, mitochondria, the endoplasmic reticulum, even arteries, veins, and capillaries—indeed any and all parts of living things—can be considered adaptive in their own right is either metaphorically, as if they were adaptations, or as inanimate objects, guarding against their own material destruction by environmental forces. This is because the adaptive properties we attribute to them only exist in context and in action. And the context and actions we are referring to are those of whole organisms, not their parts.

Things are somewhat murkier about adaptive properties for populations of organisms. It is true that in their host, populations can be advantaged in ways that single organisms cannot, and there are many examples of this. Most obviously, in our own world, we organize into groups—families, tribes, nations—for mutual protection against nature's diverse challenges; for example, attack by other groups or food acquisition. There can be properties of groups and power in numbers, and we certainly can call such properties and advantages "adaptations" of populations of organisms if we wish (as population genetics frequently does).

But from the vantage point of a single organism, these rewards would seem to be positive environmental factors, not adaptations. It is merely that the environmental factor is an organism or group of organisms of the same species, of our "conspecifics"—not of a different species. Why should one (different species) be called an environmental factor and the other (the same species) an adaptation? And while it is true that *species* tend to disappear when selection does not act in their favor, this usually occurs bit by bit, one *organism* at a time, not in some unifying cataclysmic event that destroys the whole species simultaneously. The significance of this is that the adaptive properties of populations are divisible by their individual members; they are not a phenomenon of the group as a whole, though they often appear to be.

But there is a most important argument to the contrary. While a central act of evolution is reproduction, its value for the participants' own survival is often minimal, and it is sometimes downright disadvantageous, even at times leading to death. From the standpoint of evolution, the outcome of reproduction is the propagation of a species, a particular family line, not the survival of its individual participants. That is, selection, in this case, appears to occur at a supraorganismal level.

## The Chicken and the Egg

Though we cannot attribute adaptive properties to some particular material aspect of an organism, to the sum of its materiality, or even to it being organized in a specific way, when the organism endures, so *pari passu* do its material features, from its DNA to its organization. This means that as the adaptive features of life evolved, so did its material incarnation. Critically, and this is the point, it was not the other way around.

We cannot say that as life's material features evolved, so did its adaptations, even though this seems to be exactly what happened. Certainly, the material features of life came into being *before* its adaptive properties. Each and every material necessity foreshadowed the emergence of the related adaptive sufficiency. How could it have been otherwise?

And yet it is the adaptive features of the organism, not its material incarnation that are the products of evolution. As things happened, the adaptive properties of life dragged its material being along for the ride, not the other way around. *Yes, life's adaptive properties* emerged *from a particular preexisting material incarnation, but that incarnation only* evolved *because of the adaptive properties that emerged. They were the driving force.*

And so, it is our adaptations that endow us with fitness and are the subject of natural selection, not our particular material incarnation. And though the unique materiality of living things is the foundation upon which the evolution of the species came to pass, it did not determine either its nature or path. That was determined by the actions of natural selection on life's putative adaptive properties. And it was as a consequence of these properties that the organism survived or did not survive, produced adequate numbers of offspring or did not. And of course if it did not survive, if its reproductive abilities were inadequate, then neither did its material incarnation. Conversely, if it was fit, if its qualities were sufficiently adaptive, then its particular material incarnation persisted, though dependently, in it and its offspring.

## Perfusion, One More Time

In case the distinction between life's materiality and its adaptations is not yet clear, let's look at a few other examples to make things as plain as we can. Let's begin once again with vascular perfusion. When I said that vascular perfusion is an adaptation that provides oxygen to our cells and tissues, I was not referring to the material things that serve this function, to blood vessels or the heart, or even to the physical process of perfusion. These are the characteristics that allow for the adaptation—they are necessary for it, but they are not it.

Consider two theoretical fictions. Imagine that I could somehow perfuse my cells and tissues with blood *without providing them with the oxygen they need,* without fulfilling the adaptive function of this act. If there were some way to have circulation continue unabated in the absence of oxygen, the structures of the circulatory system—the heart and blood vessels—would still be in place, and perfusion through the vasculature would still occur, but it would no longer serve this adaptive function. It would not provide my cells and tissues with needed oxygen. In this case, we would have the curiosity of a dead body (or a nonliving body) that circulates blood.

Or turning things around, imagine if you will that I could somehow suspend the demands placed on me by the environment, say by increasing the amount of oxygen in air or altering the laws of diffusion and geometry, to allow me to meet my need for oxygen by simple diffusion through the skin. If I could do this, then, even though the anatomical structures of the cardiovascular system would still be in place and my tissues and organs would still be perfused with oxygen-containing blood, perfusion would no longer be an adaptation to this need. It would no longer offer any selective advantage because I could reliably obtain the oxygen I need in another way. In this case, if my heart failed, I would still be alive.

The *adaptation* of vascular perfusion refers to the property that permits the body to meet its need for oxygen in face of the environmental challenge (too low a concentration of oxygen in air to diffuse through the body otherwise). As just explained, if vascular perfusion occurred without fulfilling this need or if the need could be met in another way, then it would not be an adaptation, at least not the adaptation I have described.

What this means is that perfusion of our tissues and organs with blood is not adaptive *inherently,* simply because it occurs, nor are the heart and blood vessels adaptations merely because of their existence. The manifestation of the adaptation depends on the presence of particular environmental challenges that perfusion helps the organism overcome. The adaptation is contingent upon these circumstances. It is not inherent to any particular anatomy, chemistry, or physics.

## Two Other Examples

In a similar way, each and every chemical and physical step between DNA (and its code) and the production of a protein only provides the *basis* for adaptations. They are not adaptive in and of themselves, either singly or together. Though this sequence of events is critical, absolutely necessary, for life, each step involved in "transcribing" the code from DNA to messenger RNA (mRNA) and "translating" or synthesizing the protein from mRNA is nothing more than a physical and chemical occurrence.

This may seem peculiar since there can be no life in their absence—we must have our proteins—but as we have learned, necessity, however absolute, does not automatically yield sufficiency. Its necessity does not make protein production an adaptive phenomenon. Not DNA, nor its code, nor its transcription or translation, nor any or all of the complex steps involved in these events, indicate life's presence. Indeed, under appropriate conditions, we can manufacture proteins quite nicely in a test tube outside the cell or organism as chemical and physical events of inanimate objects.

Consider one final example. Imagine that a changed environmental circumstance calls for the manufacture of a new protein. Say a colony of bacteria has lost access to a needed foodstuff and it is now required to use a different nutrient. In one way or another, the cell senses the need and produces (induction) the required protein. Let us say that it is a "transport" protein embedded in the cell's enclosing membrane that "carries" the new nutrient into the cell. This new material incarnation certainly seems like an adaptation to need. After all, the new protein allows the bacterium access to the new foodstuff and is manufactured in response to a perceived environmental shortage. Yet it is not.

We can understand why by imagining an untoward circumstance. Suppose that this strategy has worked well for this bacterial species for more generations than we can count, and has allowed it to survive the periodic loss of its usual nutrients, but this time there is a substance in the medium—an inhibitor—that binds to the transporting protein and prevents it from carrying the alternative nutrient into the cell. It *inhibits* its entrance into the cell. As a result, and lacking other means of obtaining food (or the ability to undergo sporulation), the bacterial colony will starve to death. Indeed, if by some chance this turned out to be the species' only habitat, it would become extinct.

But notice that the inhibitor did not affect the *presence* of the transporting protein, just its effectiveness. The protein was still being produced; it was still being placed in the membrane. Indeed, the cell would continue to produce it until it lacked the energy to do so. It might even sense the need for more, since it wasn't doing its job, and manufacture more.

However despite the protein's presence, because the foodstuff could not enter the cell, there was no adaptation to need. The adaptive property did not reside in the material elements of the system—not in the protein, nor in the mechanisms that produced it. It was found, and found only, in the ability of the cell, the whole cell, the whole organism to use the new foodstuff. If it could not do so, then there was no adaptation, however much transport protein was manufactured. Consequently, the bacterium died even though it produced exactly what was required to sustain life.

So to truly understand the role of material objects in life, we have to look beyond their material properties, beyond their structure, their chemical reactions and interactions, beyond their anatomy and physical character otherwise, and even beyond their broader collective actions, to the adaptive properties that emerge from it all. It is in these unique characteristics that life resides. For all the importance, necessity, and inherence of our material embodiment, not chemistry, physics, anatomy, nor all of them together give rise to life in and of themselves, intrinsically.

## Taking Action

As suggested earlier, inanimate objects also have adaptive capacities. For example, the rate at which a rock is eroded by the wind and rain depends on its hardness, an emergent property of the materials that comprise it. A rock made of granite is harder and hence less easily eroded than one made of clay, and its relative hardness can be considered an adaptive quality. How do such adaptations differ from the biological ones we have been talking about? Why do some adaptations to environmental circumstance imbue objects with life, while others do not?

The difference lies in the *submissive* or *passive* nature of the inanimate object and the *reactive* or *active* one of the living thing. From the rock's point of view, its engagement with the environment is entirely and unconditionally passive, and is so at all times, under all circumstances. It can only comply with the forces that are imposed on it. It may do so more or less readily; that is, it may be more or less adaptive (harder or softer), but it must comply. It can only display passive resistance. As such, while one rock may be less easily eroded than another in the face of a particular physical corrosive force, nonetheless, it is necessarily eroded to one degree or another by that force, however slightly.

But there are also inanimate substances that are not inert, that participate in chemical reactions. Aren't they reactive, just like living things? Yes, but their reactivity is of a fundamentally different sort, and this is key. Unlike living things, *chemical reactions cannot be said to be adaptive because even though mass is conserved, the substance itself is not.* Something new, something changed, modified, elementally different comes into being that takes the place of the original substance. And since the sole purpose of an adaptation is conservation of the object, chemical reactions, at least as isolated phenomena, by their very nature are the opposite of adaptations. The object, the substance is not only *not* conserved, it is destroyed and replaced with something else.

Living things on the other hand can be both reactive *and* self-maintaining. Even when their substance is changed by the forces imposed on them—for example, when a cell's complement of proteins is changed—unlike a simple chemical change, this change is designed to promote the whole object's, not a particular chemical's, survival. It is in the concordance of reaction and self-maintenance that events in the world of the living are adaptive. It is this reaction toward self-maintenance that makes them adaptive.

And so, living things do not merely comply with the physical forces that are imposed on them, acquiescing to the path of least resistance, but engage the environment both physically and chemically in an active fashion, transcending their material antecedents in the process. They are able to take *action* to improve their situation. In so doing, they *react* to what they face. And remarkably, their actions and reactions may not only better their condition, they may allow them to maintain themselves wholly unaltered by a particular threat, to be unscathed by it. Entropy can be held at bay. We can literally dodge bullets, escape a predator, find a protective haven from the wind and rain, or engage in a myriad of other actions that are not only designed to advance our cause, but to wholly prevent degradative or destructive change.

Still some biological adaptations seem to be passive features of organisms, no different than the hardness of a rock. For example, adaptations that involve coloration or the shapes of beaks seem like hardness—passive material features of objects. But as it turns out, in relation to their adaptive function, they are not. Their passivity is only apparent. This is because biological adaptations related to color and shape are not to be found in color or shape themselves, but in the uses to which they are put, in the actions in which they participate—for example, in camouflage or feeding, adaptive devices designed to secure survival.

We can understand the active nature of such seemingly passive characteristics by turning to death once again. Though the shape of beaks, the color of feathers, and many other features of living things remain unchanged at life's end, they no longer serve the adaptive purposes they once did. That is, they no longer provide an advantage in the animal's struggle for survival simply because that struggle has ended. Now all the object and its various features can do is passively comply with the forces that are imposed on them. Like rocks or a stream finding its way down the mountain, they follow the physical and chemical path of least resistance that nature has lain out. And if they are fragile structures, they rapidly decompose and decay.

With these ideas in hand, we can imagine life's beginnings in terms of adaptations. We can say that life itself came into being with the first active adaptation, in all likelihood an emergent property of a simple chemical or mechanical process, like the transport of a nutrient into a bacterium. Today, the adaptive actions of living things are immensely varied. As with early life, they can be based directly on chemical and physical processes, but they may also be found at a great remove from them, as in the many adaptations of human mental processes.

In this discussion I have used the words "action" and "reaction" as synonyms, but of course they are not. Though a reaction is necessarily an action, at least in biology an action need not be a direct reaction to something. Though actions must have their causes—they must react to something—in the case of living things, those causes may not be a reaction to an immediate or personally *experienced* need. And while most adaptive properties in life are indeed reactions to extant environmental circumstances, remarkably, some, particularly among humans, seem preemptive, not reactive. For example, it is not necessary for us to personally experience the consequences of low atmospheric pressure to design airplanes with pressurized cabins. There is certainly cause for our action—knowledge of the atmosphere, the experience of others—but it is not our own personal experience of the phenomenon.

In any event, with or without choice, with or without conscious intervention, reactive or not, all adaptations in living things are actions. To advance its survival, the organism takes action. It does not merely acquiesce. It is in this unique (active adaptive) engagement with the world that we find the *sine qua non*, the essential condition of aliveness.

# CHAPTER 9

# Spandrels and Other Irrelevancies

## *Are There Sufficient Properties of Life That Are Not Adaptations?*

I celebrate myself, and sing myself,
And what I assume you shall assume,
For every atom belonging to me as good belongs to you.
I loafe and invite my soul,
I lean and loafe at my ease observing a spear of summer grass.
**—WALT WHITMAN, *SONG OF MYSELF*, 1855**

IT MAY HAVE occurred to you that when you die it is not only your adaptive actions that are lost, but all of your actions, save none, however insignificant, however lacking in adaptive value. With death, all the everyday contrivances of life disappear.

Needless to say, you cannot get out of bed in the morning, take a shower or bathe, shave, have orange juice, eat cereal or toast, read the newspaper, put on a pair of pants, a dress, socks or stockings, tie your shoe laces, start the car, or take a bus. Nor can you buy tickets to a ball game, opera, play, symphony, rock concert, or lecture; purchase a stylish new outfit; eat sushi, a burger, steak, or tofu; drink a vintage wine or a cheap beer; read a great novel, or a trashy mystery; watch TV—the history channel or the latest silly sitcom; or choose among any of these.

Neither can you talk to a friend about what interests you, about your life, his or her life, about anyone's life, about sports, politics, history, art, philosophy, the weather, or anything else however trifling. Nor can you contemplate any of these, or for that matter any subject whatsoever, however inconsequential. Finally, you do not feel emotions—not anger, sadness, happiness, anxiety, concern, hope, pessimism, or frustration—however unimportant or fleeting such feelings may be.

Each of us can make an endless list of the trivia of our lives that end with death that seem to serve no adaptive purpose whatsoever. With our demise, all the personal and social activities of our lives, however nugatory, are gone, and like our adaptive properties, gone instantaneously. Are they not also signifiers of life? And does this not mean that life has sufficient causes that have no adaptive content? If so, what does this say about the equation I have made between life and its adaptations? Yes, they vanish with death, but then so do so many other activities.

## The Nonadaptive Properties of Life

As said, without variety, there can be no evolution. To Darwin, the variations central to evolution were the small differences between individual organisms of the same species. A particular variation might in some small way foster an organism's chance of survival and enhance its capacity to propagate relative to others of its own kind. Of course a variation might also hinder its ability to survive life's vicissitudes and reproduce or be without effect.

Whatever the variation, whether large or small, these are the only choices that nature offers—positive, negative, or without effect. We say that variations that improve an organism's odds of surviving nature's challenges (and having offspring) have adaptive value and that it is the task of natural selection to sort such helpful, that is, adaptive variations from harmful ones by favoring organisms that express the former and eliminating (eventually) those that exhibit the latter. Heritable neutral variations, neither advantageous nor adverse, are not acted upon by natural selection. Having come into being, they simply remain in place.

Conforming to classical Darwinian theory, only features that are *favored* by natural selection *evolve*. Accordingly, if harmful variations ultimately die out and neutral ones do not evolve, then all features of living things—the products of billions of years of evolution—should serve adaptive purposes of one sort or another. But given what I have just said about the triviality of the "everyday contrivances of life," this does not appear to be true. Some of our characteristics seem to serve no adaptive purpose whatsoever.

But is this correct? Or against appearances and however obscure, are they all adaptive in one way or another? This question is fundamental to Darwinian theory, and its discussion dates back to the publication of *On the Origin of Species*. Darwin and Wallace agreed about many things, but not this. Darwin thought that some of life's features lacked or seemed to lack adaptive value, and yet evolved. The incredible feathers of the peacock bothered him. What purpose could they possibly serve? While he eventually concluded that peacock feathers and many other seemingly meaningless external features of organisms, such as the striking colors of flowers, were probably sexual attractants, there was still much about the appearance and behavior of organisms that seemed to have no obvious adaptive function.

Contrary to Darwin, Wallace believed that all of an organism's characteristics *had* to be adaptive even if he could not identify the adaptation. He relied on the theory of natural selection itself for this conclusion. How else could they have come into being, he asked? If evolution could take place without the emergence of adaptations, then on what basis did it occur? If it was not adaptive value, then the whole edifice of evolution by natural selection that he and Darwin envisioned was in jeopardy. That is, if features could evolve without adaptive purpose, then evolution was not dependent, or at least not solely dependent, on natural selection.

Darwin's point of view was an expression of his faith in naturalist observation over theory. He saw what he saw, or more to the point, didn't see what he didn't see. Wallace on the other hand was more comfortable with the idea that purposes existed that were beyond his ability to fathom, and consequently placed the need to comport with theory ahead of observation, and the theory of evolution by natural selection required that evolution inevitably and unavoidably produce adaptive characteristics, and them alone.

## Spandrels

Today, in agreement with Darwin, we believe that characteristics can evolve and yet be without adaptive purpose. But notice that I did not say "evolve *without* adaptive purpose," but "evolve *and yet be* without adaptive purpose." There is an important difference. We do not imagine such nonadaptive traits as "freestanding." That is to say, we do not believe that they evolved in their own right as meaningless characteristics. Rather, though not adaptive themselves, their evolution is understood to be *connected* in one way or another to adaptive functions.

For example, at some time after its evolution, a feature might cease to serve the adaptive role for which it originally evolved. Perhaps the circumstance was no longer relevant, or no longer existed, or some other property had assumed responsibility for the adaptive task. Consequently, what was once an adaptation might become a neutral or even a harmful characteristic. We can say that it "devolved." The most common example of this is the wings of flightless birds—once useful, now seemingly useless.

Following the great British physiologist and geneticist J. B. S. Haldane many years earlier, Stephen Gould and his colleague, evolutionary biologist Richard Lewontin, proposed another way in which adaptationless or functionless characteristics can evolve. They can be *by-products* of the evolution of properties that serve adaptive purposes. They called such features, at least those with an anatomical incarnation, "spandrels," a term they borrowed from architecture.

In architecture, spandrels (pendentives when three-dimensional) are the roughly triangular areas (or objects that occupy the space) located directly to the left and to the right of the peak of a supporting arch, and just below the structure being supported by the arch, such as a roof. Spandrels come into being as the formal consequence of the geometric relationship between the arch and the object it is supporting, but serve no structural purpose. They support nothing; they cover nothing. They are merely the outcome of erecting structures that do.

The chin is often given as an example of a biological spandrel, the unavoidable consequence of the fusion of the left and right parts of the lower jaw or mandible. Their fusion makes the chin. Moreover, in humans, but not in our close primate relatives, the chin usually protrudes. This protrusion is called the *mental protuberance*. And like the chin itself, this difference between humans and primates is thought to be without adaptive purpose, the result of chance (though inherited) variations in the pattern of growth of the mandible of different species. The mandible is the moving part of the chewing apparatus and as such is critical for chewing, and as such certainly has an adaptive purpose, but the chin, though pointed, seems pointless. (To take another example of a useless appurtenance, the ear lobule—that little peninsula of tissue at the bottom of the external ear—seems useless, other than for holding earrings.)

## Soft Spandrels

But spandrels need not be *structures*, like the chin (or the ear lobule). They can also be *processes* or *mechanisms* and may not even have a material incarnation. For example, some of the everyday activities of life listed at the beginning of the chapter are, or seem to be, spandrels. We can call them *soft spandrels*, analogous to useless computer *soft*ware, as opposed to functionless *hard*ware (like chins). As with hard spandrels, soft spandrels owe their existence to traits—actions and activities—that serve adaptive purposes, though they themselves do not.

For example, though language has many important adaptive functions, speech can be frivolous, without apparent utility. We can say that idle speech is a soft spandrel whose existence is allowed (though unlike hard spandrels, not predetermined) by the evolution of the capacity for language and the adaptive functions it serves. Clothing fashions are another example. Not all that long ago, human beings learned the advantage of covering their hairless bodies to protect themselves from cold weather and found ways to construct clothes to realize this adaptation. But this year's fashion statement serves only style. It is a soft spandrel of couture. And though eating food is obviously critical for survival, eating a Napoleon, tiramisu, or making a reduction sauce is not. However delicious, however desirous, they are soft spandrels.

## Is Ignorance Bliss?

But there is a problem labeling a characteristic as being without adaptive value. It is trying to prove a negative. We may think it self-evident that a particular characteristic serves no adaptive purpose, but self-evident or not, this conclusion might reflect our ignorance, not our perceptiveness, and we should not take ignorance to be bliss.

In this light, is the shape of the human chin really a spandrel; is it really purposeless? As explained, at its most dependent point, it juts out as an outcropping of bone, an excrescence that contains two subsidiary features. The first is a groove or cleft that runs down the center of the protuberance. It is where the line of fusion between the two mandibular bones rests. To its left and right, on each side of the chin, there is usually a bump, a nodule of compact (dense) bone called the *mental tubercle*. Check them out on yourself—the groove and the bumps—most of us have them.

Can we imagine a function for these inherited features, for the chin with its protuberance, and its groove and bumps, as Darwin did for peacock feathers? While walking or running, I trip on an object and fall flat on my face. I try to break the fall with my hands, but fail. At the same time that my hands reach out, I arch my neck, and jut out my chin. Such falls are inevitable and frequent for a toddler who is beginning to walk or for an active child. When we fall or a child falls and the hands fail, the chin may break the fall. Would we be better served if we landed on our teeth, on our mid-to-upper face with its fragile underlying bones immediately adjacent to the brain and eyes, or the chin?

Unlike the other bones of the head, the mandible hangs from the skull, held in position by various muscles and tendons, and as such has a "soft" connection to the skull. This allows the jaw to be moved for chewing, but it also serves to cushion the force of percussive trauma. Through the chin, the jaw can absorb much of a blow's energy. Is this not an adaptive property?

It is a little more difficult to imagine adaptive purposes for the central groove and the mental tubercles, but let's try anyway. The central groove cradles the line of fusion and protects the jaw from fracture at this vulnerable point, while the mental tubercles, the most forward portion of the chin, bear the brunt of the initial contact with the ground and help direct the force equally to both sides of the mandible and away from the groove.

In addition, as my dentist reminded me, some important muscles attach in the area of the chin. For example, two muscles of the lower lip, the quadratus labii inferioris and mentalis, insert on the surface of the chin. They are muscles of expression. Contraction of the former is said to indicate irony and the latter disdain. Two important muscles are attached to the inner surface of the chin. One is concerned with movements of the tongue, the genioglossus, and the other with swallowing, the geniohyoideus. Might not the exact placement of the insertions of these muscles on bone bear on their being able to carry out their functions properly, and wouldn't this be dependent upon the bone's shape? Though our primate cousins have similar muscles, yet no protruding chin, their placement may have meanings that are uniquely human, say in regard to talking or our expressive nature. (Even the inconsequential ear lobule is part of the immensely active heat exchange system that is the rim of the outer ear.)

Similarly, we can ascribe adaptive utility to the soft spandrels listed at the beginning of the chapter. When I shower, I remove bacteria and dead skin (substrate for bacterial growth) and this prevents skin infections. Tying shoes to my feet prevents accidents as I walk. Wearing the latest styles may attract a mate. And of course, whatever my favorite cuisine, I need to eat *something*, and if I find a particular food tasty (pleasurable) and it whets my appetite, cooking it (or even ordering it) may be a meaningful adaptation to ensure adequate food intake.

In casual conversations with friends and acquaintances, I may learn something that helps me in my daily life—I may get good advice in a time of need, be enlightened about how others perceive me, be forewarned of danger, and on and on. Even watching TV or going to a rock concert or symphony may serve adaptive functions. It may reduce stress and even provide beneficial enlightenment about the human condition.

Of course, none of this is more than hypothesis, and that is just the point. What chins do or ascribing adaptive value to washing, talking, listening to music is just hypothesis. To be more than that, we must test the idea, the hypothesis. We must provide evidence for it. For example, regarding the chin, we could apply a known force to it as well as to other aspects of the face (controls) and measure its transmission to the relevant structures with carefully placed force transducers.

In any event, if a hypothesis for an adaptive function is formulated well enough, we can test it. But how about the contrary situation—how about a hypothesis that states that no adaptive function exists? How about a negative hypothesis? However circumspect we should be in considering such hypotheses, can we, having formulated them, put them to the test? The fact is that we cannot. We simply cannot devise experiments to show that a particular structure, process, or mechanism is functionless. How do you look for nothing?

In this case, all we can do is base our judgment on our inability to *imagine* an adaptive function. But a lack of imagination is not good enough if we seek scientific understanding. Moreover, we need to keep in mind that context is everything with all things evolutionary. What may be of no adaptive value in one circumstance may be adaptive in another (and counteradaptive in a third).

Not only that, but an activity may be adaptive for one *person* (not just one species) and maladaptive for another. Going to a rock concert may give me a colossal headache, while my teenage daughter or son may not only find it exhilarating, but relaxing, even reassuring (vice versa for the symphony). And if I am under stress, tired, and preoccupied, going to the symphony or a rock concert may not be palliative, but the last thing I need. Similarly, exercise at the gym is designed to improve my health, but if I am exhausted, a nap may be far more adaptive (indeed, much to my liking, exercise may be maladaptive!). The *coupe de grace* is that often we can only say "historically" (after the facts are played out) whether a particular action or activity has been adaptive, maladaptive, or has no adaptive significance whatsoever.

These problems are particularly true for hypotheses that imagine psychological, behavioral, and social adaptations. A particular feeling or behavior may have adaptive value for one individual, be of no value for another, and maladaptive for a third. Or it might be helpful in one situation, or at some particular time, but not another.

Religion presents an interesting example. Whatever its transcendent meaning, certainly its ubiquity among cultures and centrality to them suggests that religious belief is, if nothing else, an important adaptive response to our human circumstances. Yet many people today find it of no (adaptive) value to them personally, and see it not merely as a meaningless mélange of irrational attitudes, a spandrel of superstitious beliefs, but maladaptive, socially and personally destructive, something we would do well to get rid of. But then alongside this disapproving and dismissive view, we find a great many others who find it a source of great comfort, purpose, and enlightenment—a critical adaptation that helps them navigate life's troubled waters. Given this seemingly unbridgeable difference, how can we objectively classify religion? Is it adaptive, benign or malignant, an adaptation, a spandrel, or neither, or is it all of them simultaneously, and how can *that* be?

## The Unavoidability of Irrelevancies

All of this cautions circumspection in pronouncing a particular feature as being without adaptive value. As explained, just because we can't identify an adaptive purpose does not mean that one does not exist. Equally though, our ignorance does not mean that one *does* exist. And just because we *can* imagine an adaptive purpose for some feature or property, as I have done above, does not mean that it has such a purpose, even though we admittedly have no way of certifying a negative conclusion.

In any event, despite our inability to prove a negative, as a general matter—that is, without reference to unknowable specifics—there is little doubt that adaptationless features exist. Whether hard or soft, spandrels seem to be unavoidable by-products of evolution, as they are of architecture, even if we cannot unquestionably identify any particular characteristic as being one. Similarly, as said, there are characteristics that were once adaptive, but that no longer are, or at least that no longer serve the adaptive purposes for which they evolved, as in the example of the flying apparatus of birds that cannot fly. Thus in the dispute between Darwin and Wallace (one that continues to the present day), we must side with Darwin. Evolved features exist that serve no adaptive purpose.

## Irrelevant Sufficiency

But if this is true and we reject Wallace's contention that all features of living things have adaptive value, then don't we have to abandon the theory of evolution by natural selection, as he thought? Fortunately, we do not. Both spandrels and devolved adaptations come into being, and only come into being, as the consequence of the evolution of adaptive features. As said, they do not evolve *independently* of adaptive purpose. Either they evolved for an adaptive purpose that subsequently became obsolete or vanished (devolved features), or are a by-product of the evolution of adaptive properties (spandrels). Natural selection did exactly what it is supposed to do. It favored the evolution of adaptive traits. The existence of adaptationless traits is merely a side effect, an unintended consequence.

This brings us back to the question we began the chapter with: does the existence of sufficient properties that lack adaptive purpose mean that we cannot equate life with its adaptations? What we have learned is that even though such purposeless features appear to exist, life can still be equated with its adaptations because they—spandrels, devolved traits, and the like—are sufficient qualities of life *only* as a consequence of their linkage to properties that *are* or *once were* adaptive. They would not exist otherwise. As a result, our hypothesis about the relationship between adaptations and life is secure to the extent that the theory of evolution by natural selection is secure, and we need only modify it in an insubstantial way. We can say that all of life's sufficient properties are either adaptations or are nonadaptive features that in one way or another are historically linked to adaptations. With this minor modification, we can still declare that our adaptations make us alive.

# CHAPTER 10

# Life as Complexity

## *The Nature of Biological Complexity*

───────────────── ⌐⌐ ─────────────────

There is an alternative approach which states that life arises as a nearly
inevitable phase transition in complex chemical systems. Life, this hypothesis
asserts, [is] formed by the emergence of a collectively autocatalytic
system of polymers and simple chemical species.

—STUART KAUFFMAN, *THE PRINCIPLES OF*
*ORGANIZATION IN ORGANISMS*, 1992

ANOTHER POPULAR explanation today for the living state, particularly among a certain group of biological theorists, claims
complexity as life's cause. From this perspective, the complex
nature of living things is not just a product of evolution, nor is it
simply that life is the most complex phenomenon known to us,
but it is the complexity of certain physico-chemical systems that
endows them with life, that makes them alive. Like the adaptive
view of life, this explanation transcends the material incarnation
of living things, but unlike it and in line with materialist inclina-
tions, it envisions life as being intrinsic, not to certain matter, but
to a particular complex physical embodiment.

A theory of life based on complexity can lay claim to Aristotle's notion of life-as-organization. Though something can be
very complex without being particularly well organized and well
organized without being particularly complex, in this context the
two are synonymous in that they refer to different aspects of
the same thing. We can say that complexity, *qua* organization,

introduced life to the world. Living things and their products (automobiles, computers, language) are at one and the same time the most organized and the most complex things in creation.

"Information" can be thought of in the same way. The more organized, the more complex a system, the more information it contains. Though the three words have different meanings, their source in this case is the same. The objects and processes of living things—most everything and anything from DNA to cognition and its consequences—exhibit all three at one and the same time, in one and the same feature. The root of the word "organization," *organ*-, even refers to organisms, to life itself. When we say that something is organized, we are saying that it displays properties of the sort found in living things.

Furthermore and astonishingly, aside from the information inherent to their complex embodiment (that found in DNA, proteins, etc.) and the fact that some species introduce information in their communications or communal behavior, information in our world is almost exclusively the product of one species—humans, the information-producing animal. Though it cannot be quantified, it seems that information from all other sources combined pale in significance when compared to what emanates from humans—from the sounds we incessantly emit, to our ever-increasing fund of useful and meaningless facts, data, and knowledge, to the mass and diversity of the objects we construct.

This said, two relatively simple *inanimate* states—eddies of water and clouds—provide good, easy-to-understand examples of what we mean when we say that something is complex, organized, and contains information. We call eddies and clouds *organized* because the molecules within them exhibit concerted behavior. They move together this way or that, forming *complex* curls and swirls that separate them from like molecules in the disordered medium from which they arise.

In that medium, the movement of molecules is completely disorganized—indeed, totally unpredictable or random—and as such contains no information. Whereas knowing the arrangement and behavior of some water molecules in an eddy of water or a cloud allows us to predict the arrangement and behavior of many others, as meteorologists and geophysicists do every day. This predictability reflects the *information* they contain. Yet, as complex as clouds and eddies are—think about patterns of weather and ocean currents—their level of complexity, organization, and the amount of information they contain or introduce is incomparably less than even the simplest life-form, much less human life.

## What Do We Mean When We Say A Living Thing Is Complex?

Though biologists can and often do talk if not endlessly, then often about complexity (and for very good reason given the complex nature of living things), they usually use the word in a vague and more or less nebulous way because a clear conception of what is meant by "complex" is lacking. Simply asking what it means to say that a biological object is complex, or that one is more complex than another, undoes us. We may have good, even obvious commonsense reasons for our opinion, but try as we may, we cannot specify what we have in mind in a precise and quantifiable way.

As a result, a view of life based on complexity faces enormous problems. If vagueness or an unavoidable lack of clarity is the best we can do, then how can we hope to explain life this way, at least as a matter of science? Any useful, much less convincing, explanation of complexity as life requires as a predicate a description of the phenomenon (of complexity as life) that is of sufficient clarity to enable us to distinguish, categorize, and quantify diverse levels of complexity among biological objects and processes, as well as between them and inanimate objects and processes. If we cannot do this, and as we shall see, it seems that we cannot, then we are in no better position to attribute life to its complexity than to DNA.

And yet from a mathematical and physical point of view, complexity is a relatively uncomplicated concept and fairly easy to grasp. *It is any increase in the randomness of a system from a perfect state of order.* It exists in, indeed comprises, the world between the perfectly ordered and the totally random. The less ordered (more random) a system, the more complex it is. A completely ordered assemblage, like ice or a sequence of the number one, displays a regular arrangement of its parts and is the utmost in simplicity, whereas a random system, like water vapor or random numbers, lacks any predictable arrangement among its members and is the ultimate in complexity, at this extreme lacking both organization and information.

Though we can say that life exists somewhere in the continuum between the extremes of fundamental simplicity and ultimate complexity, we can say no more. There is no *rule* or *set of rules* we can apply to specify the relative complexity of an organism or species or their subsumed parts and systems. Is a dog more complex than a lobster, a frog more complex than a tree? Are lungs more complex than the circulatory system, or vice versa, is the gastrointestinal tract more complex than the kidney, and on and on? Curiously, even though the contributions of the nervous system and stomach to our complexity are certainly not equal, exactly why they are so different is hard to specify other than descriptively. We certainly have good, indeed obvious, easy-to-justify reasons for saying that nervous system is more complex, but if there are clear and unambiguous *rules* to make such determinations, they have yet to be discovered.

But the central problem in looking for life in its complexity is the same as the one faced by biological materialists. As explained, the materialists equate our physical being with life itself by claiming an identity between the two. Yet, whether the claim is justified by analogy or metaphor, they cannot prove it true. The best they can do is provide verisimilitude. In a similar way, if complexity gives rise to life, it is not enough to just *say* that this is so. We must be able to articulate what exactly about its complex nature makes some material object living.

For example, in the epigraph at the beginning of this chapter, Stuart Kauffman says that some believe that life first arose with the coming into being of certain complex physico-chemical systems, but they do not, indeed cannot, tell us what in particular, what piece of complexity, gave this system or, for that matter, any system life. To provide an answer, all sorts of unsettled questions must be resolved. For example, at the most basic level is there a certain *kind* of complexity that produces life, or is it merely complexity in general? If the former, then what is it, and if the latter, then what is it about it? And how complex must an object be to be alive? As noted, is there a complexity borderline that if crossed makes an object alive? If so, what is that border? And is it possible for an object to cross the borderline and yet remain inanimate? If this is not possible, why isn't it? Is there such a thing as too much complexity for life? If so, what is too much? What is the correct number of cooks not to spoil the broth? For a coherent theory of complexity as life, these questions require stipulated and clear-cut answers, answers that we do not currently have.

## The Evolution of Complexity

One vital problem in looking for life in its complexity is seen in our understanding of life's origin. The origin of life is often imagined as an *ordering* event that occurred in a disordered medium—in all likelihood, a body of water. Favorable chemical or physical interactions between certain substances are thought to have led to their removal from their scattered state in an ancient anarchic medium, to form a separate, more ordered object, the living thing. That is, it is believed that life arose as the result of a mechanical isolating occurrence, or series of such occurrences that not only isolated some substances from their surroundings, but in so doing produced a more ordered and *therefore less complex* object.

This presents a paradox. It claims that life began as a contradiction. First, it came into being as a shift from the complex to the ordered (simple), and then it evolved, as if by some strange change of heart, in the other direction, as a shift from the simple (ordered) to the complex. That is, to evolve, life made a U-turn.

How or why such an event would occur is not clear, but since we have no way of actually knowing how life first began, perhaps we should not worry too much about it, and for purposes of this discussion, simply replace one unsupported supposition with another. We can solve the problem by assuming that instead of life arising as the consequence of an ordering episode in a disordered medium, that it came into being as the result of a disordering event in an ordered medium, such as ice, some other crystal, or an ordered amorphous material. In this case there would be no U-turn. Life would evolve just as it began, with both disorder and complexity increasing in parallel, with life's origin and its evolution following the same path from the simple to the complex. Alternatively, we can imagine that the ordering effect was not merely compounded in a layered fashion with each new addition to life's form and substance, each new ordering event being stacked on top of others, but was composed side by side, by the parallel alignment of things, introducing complexity in this way.

In any event, this raises the question of what some think is the astounding peculiarity that biological evolution ostentatiously disregarded the second law of thermodynamics. The second law states that disorder or entropy increases inexorably with time as systems move from an ordered to an increasingly disordered state. There is no physical basis for thinking that the evolution of life is an exception to this rule, and yet it is often imagined to have been such, with evolution having increased the order of things, thereby decreasing entropy.

But the same paradox applies to eddies and clouds. They arise from a disordered medium, gaining organization and information in the process. From this point of view, against the second law we can say that they are also more ordered than the liquid or vapor from which they arise. Like life's evolution, eddies and clouds are far from equilibrium. But unlike life they tend to be transient states that quickly fall back into the chaos of the disordered, only transient dislocations from equilibrium. They do not evolve into more and more stable, more and more organized structures, as with life.

In any event, life's origin aside, this is at least in part a problem of semantics. In the common parlance, the two operative terms—order and complexity—are often confused with each other and used as synonyms when they are really antonyms. If evolution had actually increased *order*, then the second law would indeed have been abrogated, but then we would be saying that evolution had increased life's *simplicity* (order), not its complexity. Yet it is our unmistakable understanding that the opposite occurred, that evolution increased the complexity of living things.

And so, though it may seem counterintuitive, as complexity increases so does *disorder*. The two are simply different names for the same phenomenon. Seen in this light, as its complexity has increased, life can be understood to be on an inescapable path to randomness, the ultimate in disorder, the ultimate in complexity, despite all appearances to the contrary. However well organized our various activities have become, the evolution of biological complexity, though unique and remarkable, is merely life's particular way of passage to the ultimate state of randomness, of disorder, in compliance with the second law of thermodynamics.

## The Complexity of Numbers

Though we can apply a general mathematical definition of complexity to life, it turns out to be useless for comparing and contrasting the complexity of different biological objects and processes, useless to establish their relative levels of complexity. Let us see why. The mathematician Gregory Chaitin explained complexity in terms of numbers theory, that is, in terms of sequences of numbers. As mentioned, a number comprised of the same numeral repeated over and over again, like 1, 1, 1, 1, 1 . . . , is completely ordered. Indeed, it defines order. At the other extreme are numbers that are comprised of sequences of numerals (say zero through nine for the base ten) that display no discernable pattern among their members. No sequence of digits repeats itself; they are all randomly positioned. These "random" numbers can only be described by enumeration, by writing them down digit by digit, however long they may be.

Between these two extremes of perfect order and complete randomness is the world of numbers that contain repeating patterns of numerals. The pattern may be short and easily deciphered, such as 1, 2, 3, 1, 2, 3, 1, 2, 3, or long and obscure, but either way, all such numbers can be reduced to or explained in terms of shorter mathematical expressions, as Chaitin has shown, by expressions that contain fewer symbols. This reductive quality defines their relative complexity, the complexity of their contained pattern. The greater the number of symbols the reduced expression contains, the more complex the number.

But not only *numbers* exist between the extremes of the perfectly ordered and the totally disordered. It is also here that we find the complexity of the physical world and of life. Like numbers themselves, the complexity of these phenomena can also be explained in terms of numbers theory. That is, the complexity of the physical world with its living things can be understood in exactly the same way as numbers. It can be comprehended as a set of reducible numbers between the extremes of perfect order and complete disorder.

Another feature of a number is its length. Does it contain ten, one hundred, a million, or a billion digits? Length is important when thinking about complexity because the potential complexity of a pattern of numerals within a number depends on its length. A number with *ten million* digits allows for patterns of far, far greater complexity than a number with just *ten* digits. That is, the reduced expression for a number with ten million digits can contain far more symbols (the measure of its complexity) than one from a number ten digits long. Notice that I said that the reduced expression for a longer number "can," or has the potential to, contain more symbols, not that it necessarily does. The sequence 1, 2, 3, 1, 2, 3 . . . is no more complex for having been repeated a million times, than ten times. And so, though not a measure of complexity itself, the length of a number sets *limits* on the potential complexity of any pattern internal to it.

## The Complexity Number

Now let us try to relate these ideas to living things. Let's picture a reducible number, call it "a complexity number" (not a complex number, but a complexity number) to focus attention on its complexity. As explained, all such numbers have a particular reducible sequence of numerals (the pattern of numerals within the number) and a specified number of digits (its length) that define its unique complexity. We can say that such a number, a particular reducible sequence, exists that describes or defines the *complexity* of each and every feature of life. Imagine such numbers for the DNA of different species, for different proteins, for the heart, for the kidney, for various features of living things in combination, and finally imagine an enormous, all-inclusive, grand complexity number for the whole organism.

A compendium of these numbers would tell us much about the complexity of life, but, alas, such a list is merely a dream. We cannot specify a complexity number for any aspect of biological systems, much less for them all. Indeed, when we try to get specific, we find that it is one thing to explain complexity in general mathematical terms, and quite another to describe it in a way that usefully illuminates the complex nature of living things. Before we see why, we should realize that there are only two sources of complexity in living things, and the first gives rise to the second. First, there is cellular organization/information. It is common to all living things and is incarnate in the living cell. Second, there is whatever distinguishes the complexity of multicellular organisms from that of cells.

## Complexity at the Level of Genes

Let us begin with modern biology's favorite number—the number of protein coding sequences in DNA molecules. From the time that the genetic code was discovered more than fifty years ago, it has been understood that these chemical sequences are the ultimate cause of the properties and traits of life, including its complexity. They are the "genes" of the classical geneticists, the underlying basis or genotype of life's various observable anatomical and functional characters—what biologists call the phenotype. And as we all hopefully learned in Biology 101, the genotype gives rise to the phenotype.

In this view, it is the number of characters that determines the complexity of a species. The more of them it has, the more complex it is. And because biological characteristics are the product of genes, and genes are sequences of DNA that code for proteins, we can also say that the more proteins its DNA contains the code for, the more characters it has, and hence the more complex the species. From this perspective, complexity is a relatively simple matter. The number of genes DNA houses determines a species complexity, and as such, knowing its complexity is merely a matter of counting genes.

Though I have not seen it stated as such, the Human Genome Project at least in part was based on this understanding. It was expected that when the genome of various species were compared, differences in their complexity would be explained by differences in the number of DNA sequences for proteins, in *their number of genes*. All we would have to do to assess their relative complexity is compare this number.

This expectation was not only unrequited, it was demolished. When compared to the incredible variety of life's forms and their apparently widely divergent levels of complexity, differences in gene number among species turned out to be so small as to be essentially nonexistent, and what differences were found were certainly not likely to account for the immense existential differences between them. The largest difference was about an order of magnitude between the number of the bacterial (prokaryotes) gene sequences (the genome) and those of plants and animals (eukaryotes). Given the tremendous phenotypic differences between bacteria and us, an order of magnitude seems a puny difference indeed.

But more telling, for all intents and purposes, differences in gene number disappear among nonbacterial species. Despite what seem to be great differences in complexity, the number of genes (again, meaning sequences of DNA that code for proteins) is roughly the same for species of widely differing phenotypes. And the differences that are seen, about twofold, do not track with our commonsense understanding of the relative complexity of species. For example, rice and mustard weed plants have more genes than humans (25 to 50 percent more), not many fewer as we might expect, and the latest estimates place the gene number of the lowly roundworm at close to parity with humans.

Given the deeply held belief that gene number can be equated with life's characters, this was a great surprise. But surprise or not, the facts seemed clear. The immense differences in the complexity of organisms *could not* be attributed to differences in gene number. However unexpected by molecular biology, this outcome is not surprising from the perspective of complexity numbers. In its terms, the complexity of our genetic makeup is a particular repeating pattern of numerals in a number that contains as many digits as we have genes. It is this *pattern*, not the number of genes that determines complexity. All the number of genes signifies is the length of the complexity number. As such, it sets *limits* on the system's complexity, but does not indicate its actual complexity.

Remember that while the *potential* complexity of a number with one hundred digits, or an organism with one hundred genes, is far less than one with ten thousand digits or ten thousand genes, this potential may or may not be realized, or more accurately, it may only be realized to one extent or another. This understanding provides a straightforward explanation for the seemingly odd observation that species with similar numbers of genes, such as rice, humans, and roundworms, appear enormously different in terms of their levels of complexity. Though they all have a roughly similar *potential* for complexity (having evolved over the same period of time), they are immensely different in regard to their *actual* complexity. In any event, whatever the common wisdom has been, we have learned that counting the number of gene sequences in DNA tells us almost nothing about the manifest complexity of particular life-forms.

However unanticipated, this realization was the real, though often-undeclared accomplishment of the Human Genome Project. But declared or not, its effect was dramatic, and biologists jettisoned a core belief of their discipline with surprising rapidity. That belief, etched in stone for over fifty years, was that one gene, one particular sequence of DNA, carries the code for one particular protein—*one gene, one protein*. Now we understood that genes could be reconfigured to produce more than one protein. Perhaps, it was thought, it was not the number of genes, but their permutations, and the variety of proteins that emerged from them on which differences in the complexity of species was based.

As such, abandoning the one gene, one protein concept that modern genetics thought exposed the parsimonious beauty of nature was not a radical conversion. It was a conservative, or more accurately, a defensive reaction to the results of the Genome Project. It was defensive because sacrificing this model not only suited the data, but it allowed retention of a far more important, more general, and more deeply held principle. That principle was that DNA-based codes for proteins are responsible for life's characters—its observable properties, its phenotype, however many protein molecules are coded for by each gene sequence. What really mattered was the understanding that DNA sequences are the genotype that underlies the phenotype—a matter for discussion some other time.

## Resemblance and the Complexity of Life

What numbers theory tells us about gene number applies to the number of any and all things. That is, complexity simply cannot be understood in terms of the number of things, whatever the things, substance, object, or type of event. For organisms— whether we are talking about DNA, proteins, biological chemicals in general, the number of chemical reactions, the number of cells, or the number of anatomic structures within cells or organisms—number is not a measure of complexity, only an expression of its possibility.

Indeed, trying to account for the great phenotypic diversity of life by means of differences in the substituent substances, objects, and physico-chemical properties of different life-forms has taught us, I believe, quite unremarkably that however great their phenotypic diversity, species have far more in common than they have differences. Most importantly, if one can say anything about the functions that proteins support, the reactions they catalyze and structures they form, it is that there is a considerable *resemblance* between diverse cells and species in both what they are as chemical entities and what they do. Though the details of cellular chemistry differ from species to species, as best we can tell, the similarities greatly outweigh the differences. Structural resemblances among proteins naturally suggest the same or similar functions. If anything has been conserved by evolution, it is life's chemistry, especially its protein chemistry. Just as it takes being anesthetized not to notice the great diversity in the appearance of organisms, one has to be equally inchoate not to see the enormous *similarities*—similarities as great as species are otherwise diverse—in their chemistry.

That is to say, the biochemistry of life based on proteins and their actions is much the same among diverse species. If twentieth-century biochemistry has taught us anything, it is that life's diverse forms share a common chemistry. Whatever organisms look like, however they act, whatever their phenotype, without a shadow of doubt they are built on a common chemistry. And that chemistry—the chemistry of organic synthesis and degradation, of energy metabolism, etc.—is the chemistry of proteins. Though we cannot yet provide a molecule by molecule account of proteins as we can for genes, complexity numbers tell us that there is no basis for thinking that differences in the number of *proteins* in different cells and species explains life's complex landscape, any more than the number of gene sequences in DNA.

The same can be said for cellular or microscopic anatomy. Bacteria and protozoa aside, similarities in the structures found within cells, not differences, are the rule. Cells from widely divergent species display the same basic anatomical characteristics—nuclei, mitochondria (chloroplasts in plants), the Golgi apparatus, the endoplasmic reticulum, an enclosing membrane, microtubules and microfilaments, and so on—that carry out the same or similar tasks. Where anatomic diversity is seen, with the exception of stem cells, it reflects the fact that different cell types carry out different, specific, and narrow band tasks (differentiated cells) such as contraction, motility, the absorption of nutrients, the excretion of waste products, secretion, and communication. But whether in the same or a different species, to the best of our knowledge such differences are not in themselves correlated to a hierarchy of complexity. They merely reflect the anatomy that underlies a particular activity—however complex or simple it may be.

In the final analysis, and with an exception of sorts for bacteria and protozoa, whether we are talking about DNA, proteins, or cellular anatomy—that is, about complexity at the cellular level—as remarkably complex as the biological cell is, its complexity appears to be far more a reflection of *similarities* between disparate objects than differences. On this basis, there is no reason for thinking that the "complexity numbers" of cells—either their length (the number of objects or processes) or reducible pattern (their actual complexity)—vary much among different cell types. To grasp the enormous, seemingly self-evident differences in the complexity of *species* and their substituent elements, we have to look elsewhere: we have to look beyond the cell.

## *Beyond the Cell*

As explained, there is only one other place we can look. It is at the groups of cells that form multicellular organisms. Despite the caveat about the futility of counting, the most obvious source of complexity in this larger world is the number of cells that comprise the organism. Indeed, it may seem obvious that the more cells an organism has, the more complex it is likely to be. Certainly, single-celled species are the simplest, while large mammals, comprised of trillions of cells, are the most complex. Multicellular organisms with small numbers of cells, such as the roundworm with a few hundred cells, appear less complex than species with large numbers of cells, say crabs or mice.

But this inference from cell number to complexity does not work either. There is no more of a connection between the number of cells and the complexity of an organism than between its complexity and the number of DNA coding sequences. While the cell-poor roundworm can only be as complex as its number of cells allows, the most cell-laden organism is not necessarily, nor even likely, to be the most complex. In fact, many organisms that are abundantly endowed with cells are relatively simple. For instance, the sponge, near the bottom of the complexity hierarchy, usually contains many more cells than a compact but very complex insect. Nor does the enormity of cell number of the elephant or whale make them more complicated than we relatively cell-poor humans.

The same conclusion applies to the tissues and organs of animals. Even though species that have tissues and organs appear more complex than those that lack them, their number is not a useful guide to complexity. It is not much different among species that appear to exhibit enormous differences in complexity otherwise. In any event and in the final analysis, whether it is the number of genes, proteins, cells, tissues, organs, or all of them combined, if counting is our only tool in the search for understanding the complex nature of life, we are lost.

As numbers theory explains, if we hope to specify the complexity of an organism or some feature of it, we must look beyond the counting of parts to the number of symbols that remain after a complete mathematical reduction of complexity numbers. If we could determine this quantity for different organisms and their various aspects, we would have the means to characterize their complexity relationships. But there seems no way of doing this. Though such values exist, we have no way of knowing what they are. This is the sad fact of our current state of understanding, and it appears insurmountable. At least for the present, complexity numbers only exist in theory. This does not mean that they are indeterminate, merely that we are unable to determine them and, as a consequence, are unable to assess life's complexity in other than ordinary, commonsense terms.

## The Ultimate Level of Organization

This said, having completely discounted the enumeration of parts as a means of assessing complexity, we must admit to a powerful counterexample, an example in which number appears central. There is one aspect of animal life for which counting parts provides a meaningful comparison of complexity, or so it seems. Amazingly, it is for the most complex feature of living things: the nervous system of animals.

Despite its immense entanglements, it seems that differences in the complexity of the nervous system can be understood in terms of differences in the number of nerve cells (neurons) and the number of connections between them. The more neurons there are and the greater the number of connections, the more complex the nervous system. This fact does not merely present an exception to the negative rule about counting that I have just laid out, it completely undermines it. This is because the nervous system is the major source of complexity in animals, indeed the major source of complexity period, bar none. Indeed, in comparing the complexity of species, it can be said—fairly, I believe—that the more complex the nervous system, the more complex the species. All other differences are trivial.

And so, contrary to what I have said, it seems that in this most important case, complexity *is* in fact a matter of counting. Yet, despite appearances, once again this is not the case. Even with numerical precision, attempting to measure the complexity of the nervous system this way only provides us with an impressionistic notion. However odd it may seem, counting the number of nerve cells and the connections between them does not provide a quantitative assessment of the relative complexity of the nervous system any more than counting genes does for the organism. There are two reasons for this and they are formidable:

▸ First, knowing the number of nerve cells and the number of

connections between them is not enough to fully characterize the complexity of networks of neurons, and

▸ Second, the nervous system is not merely a network of
neurons.

Ultimately, the complexity of such networks is characterized
not by the number of cells and their connections, but by the *patterns* in which these connections are made. It is here that we find
the reducible complexity number. Otherwise, as with DNA, proteins, etc., knowing the number of neurons and the number of
connections between them tells us the *length* of the complexity
number, or the potential complexity of the system, not its *actual*
complexity.

The second reason is even more compelling. Even if we could
count the number of cells and their connections and extract a
complexity number from the complex pattern of connections,
we would have only learned about the complexity of the neural
network, *not* the complexity of what the nervous system actually *does*.

And it is the complexity of what it does, it is the action that
is relevant if we wish to understand the complexity of the nervous system in functional terms. Even if it seems intuitively obvious that more complex neural networks underlie more complex
acts, to say that more complex networks *indicate* or *demonstrate*
more complex acts, and that we can form a hierarchy of such acts
based on a hierarchy of networks, is to confuse causes and consequences, substance and action, and means and results. It is to
confuse the complexity of the nervous system as a physical entity
with the complexity of what it *does*.

Though the two are of course related—there can be no sufficiency of action without the attendant necessity of substance—
we cannot infer the former from the latter. We cannot say that
what I do is more complicated because *it*—the neural network—
is more complicated. This is like saying that my clothing keeps
me warm, not because my body temperature does not fall when
I wear it, not because I do not shiver when I wear it, not because
I do not feel cold when I wear it, but that it keeps me warm

because it is clothing. Once again, an identity is assumed when proof is required.

The fact is that to understand the functional complexity of the nervous system, we must gain a sense of the complexity of what it does that is independent of the number of neurons, the number of connections, and the patterns formed among them. Otherwise all we can say is that species with more neurons, more connections, more complex patterns of connections have more neurons, more connections, and more complex patterns of connections—*however simple or complex the actions of the nervous system may be.*

This may seem obviously wrongheaded. Doesn't it make sense that the more complex the network, the more complex the resulting action? And doesn't it seem very likely that the second follows directly from the first? But this is just not the case. Being able to independently judge the complexity of particular outputs of the nervous system—our actions and thoughts—is the predicate for correlating the number of neurons and their connectivity to the real world of life's complexity, not the other way around. That is, if it is true that more complex networks underlie more complex actions, and that there is a correlation of some sort between the two, then the correlation can only be made if we know the complexity of the action independently. If we argue that we cannot determine whether a particular action is more or less complex *except* in terms of the complexity of the underlying web of neurons, then we are claiming circularly and without proof that the more complex the web, the more complex the action.

But as with everything else we have discussed about complexity, there seems no way to estimate the complexity of acts, of obtaining their complexity numbers. For the simple case of bodily movements, we might conclude that it makes estimable sense that more neurons are needed for more complex movements. But if this is so, we must first establish how complex a particular movement is in its own right or relative to others, and *then* look at the number of neurons, etc. But this assessment cannot be made, and if we cannot do this for something as straightforward as motion, how can we ever hope to assess the relative complexity of the manifold processes of the nervous system involved in such complex tasks as obtaining food or avoiding becoming food?

Moreover, even if we could establish that the complexity of a neural net was related to a particular outcome, we would still have to establish the nature of the transformation between the two. If they commute, we must show how. Assuming that a correlation exists, how does the complexity of the network scale with the complexity of the action?

In all, prospects for such understanding seem pretty dim. How can we correlate the presence or absence of, say, the ability to do calculus to the relative complexity of the nervous system? What is the calculus of calculus? Are small invertebrates who cannot perform calculus (at least as a conscious act) less complex *as a result* than humans who can, and if so, by how much? And are humans who cannot do calculus less complex than those who can, and if so, again by how much? And how do we determine the complexity of the presumably calculus-free, but nerve cell–laden whale in comparison to the calculus-enriched, neuron-poor human?

And so, to think complex thoughts and do complex things is not the same as, or equivalent to, the complexity of a web of neurons, and there is no obvious way of mapping one onto the other. This difficulty is not merely something to note, it is of great consequence. For well over a century, science has believed that the nervous system works much like an inordinately complicated electrical circuit. That is, it functions through the agency of multitudinous component circuits formed by connections between nerve cells that carry electrical charge to and fro. Though other models have been suggested from time to time for memory and consciousness, the idea that neuronal circuits explain brain function has not merely dominated thinking about the subject almost from the time that electricity itself was discovered, but it has pretty much stood alone, without a substantive alternative.

Yet science enjoins us not to accept models by default. They must prove their worth, their applicability. But how can we do what we do not know how to do? That is, we do not know how to correlate the complexity of the various actions of the nervous system to parameters related to its circuitry (such as the number of cells and their connectivity), while it is only by doing so that we can say that its electrical circuitry explains how the nervous system works. And the hope of specifying the number of neurons, the number of connections, and the pattern of connections that give rise to even the most basic thoughts such as "I am hungry and want food," much less those that led Einstein to invent his theories or Mozart his music, seems far, far beyond us.

Nonetheless, despite such intractable difficulties, the fact that the mere number of parts plays a key role in the complexity of the nervous system when it does not seem to elsewhere is a significant fact. It is a reflection of the nervous system's unique nature, that is to say, its organization. The character of the work performed by other tissues or organs is based primarily on what cells do, not their number or how they are connected to each other. Though coordination of one sort or another usually occurs and can even be critical—for example, cells in the kidney must exist in a particular juxtaposition to each other to process urine— coordination in *its own right* is not the *purpose* of the cell. It is a means to some other end (making urine). For the nervous system, it can be said that its enduring, if not its only, purpose is coordination. It is both means and end.

In any event, whatever complexity has been introduced into the world by the living cell with its DNA and proteins, whatever complexity is embodied in the heart, or in any and all of the tissues, organs, and systems of the body, pales into insignificance when compared to that introduced by the nervous system. As already suggested, it is not merely the *sine qua non* of complexity, its greatest effulgence, it is its principal source, period. That is why the nervous system is the *ultimate level of organization*, and perhaps, given its diverse and flexible array of adaptive functions, the ultimate expression of life.

## Looking for Answers in Development

For some scientists the key to understanding variations in the complexity of living things lay in the study of embryology—that is, in learning how mature organisms develop from a fertilized egg. This belief is based on the knowledge that complex multicellular organisms, however unlike they are otherwise, develop from these seemingly interchangeable, almost indistinguishable objects. Since Aristotle, scholars have thought that if we could understand how such similar things grow into such different creatures, we would understand the basis for the differences between them.

In recent times this effort has focused on the regulation of gene expression—that is, on where and when different proteins are manufactured. By initiating, ending, and altering the production of different proteins in response to various environmental and local stimuli during development, the regulation of gene expression is thought to ensure that the various features of the cell and organism are put in place when and where they are needed. There are two main events in the creation of an organism from an egg: *cell division* and the *differentiation* of cells into particular types. Through differentiation the regulation of gene expression gives rise to cells of various and sundry sorts that self-segregate to form the body's disparate tissues and organs, such as the liver and heart, each with its own unique anatomy and physiology. At the same time it is by means of cell division that the organism creates in addition to mass its unique topology—its shape and appearance.

In considering these two events, we must be careful to distinguish between their initial causes (what initiates them) and their proximate causes (the events and mechanisms that actually produce a particular material embodiment). The regulation of gene expression is an initial cause, not a proximate cause, and as such, and however critical, produces nothing in and of itself. To use a theatrical analogy, it sets the stage, commences and puts an end to each scene, raises and lowers the curtain, dims and raises the houselights, puts the correct scenery and props in place at the right time, makes sure that the actors are wearing the right costumes, prompts them when to enter and when to exit—stage left or stage right—where to stand, and only then, to utter what the script says. For the play to be realized, these are critical jobs, but they neither write nor perform it.

Agreed, you might say, but can't we fairly argue that we already have the script (DNA) and know the players (proteins)? Isn't the timing and location (which cells are involved) of events all that is missing? If that can be established, won't the production of life simply unfold? Yet, and this is a radical thought, how clear is it that DNA provides the script, and that proteins are really the players? DNA—grandiloquently and reverently referred to as The Book of Life, as if it were the atheist's alternative to the Bible—does not look like a script at all. It offers no narrative theme, no story, only characters. It seems more like a list of players, proteins, along with instructions for when and where they should be used. It tells us little or nothing about the play's story, about the form and function of life.

And what if proteins are not really the players, but merely pencils, pens, and paper with and on which the play is written? Certainly they would be many types of pens, pencils, and papers, but perhaps that is all they are? Could we have misidentified amazing pens, pencils, and papers for actors? Doesn't it make some sense that the actors are not molecules, but processes of one sort or another that by their very nature "act"?

In any event, however important for life's genesis, however central to its complexity, gene regulation merely signals events, it does not carry them out. The regulation of gene expression just does not explain differences in our form or function, or for that matter, our complexity. In this regard it has recently been appreciated that a family of genes called the Hox genes thought to be critical for development are the *same* in organisms of widely divergent appearance. Anyway, in the final analysis and despite all of the accomplishments of molecular biology, we remain pretty much in the dark about how the actual form of the body, so diverse among species, and so readily distinguishable among individuals, comes into being, becomes the particular thing itself.

## Complexity and Adaptations

After all is said and done, we are unable, except in the most abstract terms, to characterize the complexity that is life. We simply do not have the intellectual tools. We cannot specify what about its complex nature introduces life, beyond claiming an identity between the two. Given this, and with all due respect to the important and interesting ideas of complexity theorists and awe at the complexity of life, the idea that complexity is life's endowing feature must be rejected. There is simply no basis for drawing such a conclusion.

But if this is true, then in what way *does* complexity fit into our understanding of life's nature? To answer this question, we must ask about its relation to life's adaptive properties. As said, adaptations are all about an organism's fitness, about its ability to survive, but in this regard complexity is an uncertain business. While increases in complexity may enhance fitness, they may also diminish it or be without effect. Certainly, the most complex system is not necessarily, nor even likely to be, the most fit, efficient, or effective. Too many cooks really do spoil the broth.

This said, adaptations and complexity are closely related phenomena. Whether we consider complexity in general terms or in reference to a particular thing, the complex nature of an organism is like any other of its characteristics—part and parcel of the flow of evolution by natural selection—and as such, is continuously challenged by its demands. The relative complexity of a characteristic or how that complexity arises provides no exemption from this trial.

It is in this way that the complex phenomena of living things, such as the circulatory system, evolved. Their complexity is like any other adaptive asset. It is beneficial, because it has been selected. Complexity theorists refer to life being a "complex adaptive system," and place the emphasis on the word "complex," using "adaptive" much like Koshland to connote control or regulation, not Darwinian adaptation. Though the phrase is apt, the emphasis belongs on the word "adaptive," not "complex," and as Darwin understood it, not in cybernetic terms. Life is indeed a complex adaptive system whose complexity is embodied in the adaptive processes and mechanisms that have been vetted by natural selection over the course of its evolution.

CHAPTER 11

# Harmony

## *The Concordance of Life and Its Science*

———— 〰 ————

The frog, as a representative amphibian, has not acquired the means of
preventing free evaporation of water from his body, nor has he an
effective regulation of his temperature. In consequence he soon
dries up if he leaves his home pool, and when cold weather
comes he must sink to its muddy bottom and spend
the winter in sluggish numbness.

**—WALTER CANNON, *THE WISDOM OF THE BODY*, 1932**


AND SO, for better or worse, the case has been made. It is in
our adaptive abilities, not our material embodiment or complexity, that life's essence is to be found. Neither the necessity of
our material incarnation nor the complexity of our being belies
the fact that they are insufficient to give rise to life. Only in the
presence of its adaptive qualities are life's sufficient properties
revealed.

Having drawn this conclusion, we must ask whether science
can develop an inclusive grasp of adaptations as disparate as vascular perfusion and speech not merely in broad conceptual terms,
but with qualitative and quantitative precision and particularity,
making use of the reductive physical, chemical, and mathematical methods of materialist science. Or if this cannot be done, then
we must ask why. In this final chapter of *Life beyond Molecules
and Genes*, we consider the possibilities and limitations of such
a research program.

## Adaptations

As discussed, biological adaptations are grounded in a wide
range of diverse anatomical structures—seen most obviously in
the remarkably varied appearance of species—as well as in many
distinctive chemical, physical, and physiological processes. Yet,
despite their amazing diversity, they share the following general
features:

- ► They are the central consequences of evolution.
- ► As such, the unforgiving hands of natural selection forged
  them.
- ► Reproduction aside, they enhance an organism's chances of
  survival in face of life's various exigencies.
- ► Though based on the physical and material properties of
  living things, they are not inherent to them, but are contextual, contingent, emergent, and transcendent occurrences.
  That is to say, they arise from and yet are apart from their
  source (emergent and transcendent) and take place in relation to and are dependent upon other things and actions
  (contextual and contingent).
- ► They are only expressed in the context of and contingent
  upon an object's particular interactions with its *environment* at specific places and times in distinctive and sometimes singular circumstances. *They do not exist outside
  these circumstances.* We can say the same of life, and this is
  not coincidental, but causative.
- ► They are properties of whole organisms and only whole
  organisms. They cannot be found in any of their parts in
  and of themselves, however important the part may be to
  the whole.
- ► Along with associated spandrels, devolved adaptations, and
  the like, they are invariably and exclusively the sufficient
  properties of life. There are no others.
- ► As such, it is their presence that makes us alive and that distinguishes us as a matter of kind, not merely in our particulars but fundamentally from all other material objects.

## Evolutionary Progress

If, as Darwin argued, adaptive properties are the consequence of the survival of the fittest (or "fitter") organisms in accordance with the forces of natural selection, then evolution is intrinsically progressive, with progress being manifest in an abiding increase in the number and quality of adaptations. That is, if natural selection is indeed the engine of evolution, and evolution is the advancement of adaptations, then evolution is progressive *by its very nature*. As said, Darwin's theory not only predicted adaptive progress, it required it. Toward the end of the nineteenth century, as the scientific community came to accept Darwin's idea that natural selection was the *mechanism* of evolution, the *path* of evolution was understood to be progressive. Indeed, the word "evolution" became a synonym for progress.

Evolutionary progress was usually depicted in the form of a tree of life with its trunk and various boughs and branches stretching, reaching, and advancing outward and upward. At its base were the simplest and oldest life-forms—the single-celled bacteria, protozoa, and the like—while at the crown sat man, recently arrived and the most advanced species. Everything in-between was connected to show their place in life's history—what they came before and what they came after. Our lofty location was a matter of religious belief as well as everyday observation. From a religious point of view, humans—made in God's image—belonged at the top. As for what we observed, despite the many weaknesses of our species and the depredations of humans, it seemed that our language, reasoning, and manipulative skills placed us far above all others.

With our commodious brain and digital dexterity, we were undoubtedly the most *advanced* species, and our closest relatives, apes and monkeys, though far less able, were next in line, superior to all other species. Moving down life's tree and encompassing more, and putting the confusing plants aside, were warm-blooded animals who were more advanced than cold-blooded animals, and then there were the vertebrates, cold- or hot-blooded, who were more advanced than invertebrates. Next came the invertebrates, some of whom were more advanced than others, and all of whom were more advanced than the evolutionary source of them all: single-celled organisms—protozoa and bacteria—the least capable of species.

The tree of life was a powerful metaphor for biological evolution. In a single visual image, it captured its essential features:

- First and foremost, it showed that various species appeared at different, often widely separated, points in time. At the bottom of the tree were those that came earliest, whereas at the tips of its branches were the newcomers.
- Second, it related all species to each other through their common ancestry. They came into being one after another in a particular sequence that the tree's form outlined.
- Third, this sequence was branched like the tree. It was not a straight line.
- And finally, fourth, the trunk traced life's origin to a single source.

This portrayal of biological advancement reflected adaptive progress. The earliest species—the ancestors of today's bacteria—were both the most primitive, or the least advanced, and the most vulnerable, or the least adaptively endowed. As life evolved, increasingly well-adapted species came into being, and even though sooner or later almost all failed or became extinct (many in evolutionary dead ends), according to evolutionary theory, species that were less vulnerable than their ancestors and more vulnerable than their descendants served as part of the long forward march of adaptive progress. As a matter of course, in large and small steps, taken over long periods of time, with failures of all sorts littering the trail, evolution produced more advanced, adept, and increasingly adaptively endowed species.

As such, the tree of life was not merely a complex *sequence* that specified when various species appeared, but a *hierarchy* of adaptive competence. This said, science has not and in all probability will never be able to show that such a hierarchy truly exists.

Simply, we have no way of assessing the adaptive competence of creatures long gone. Still, the hierarchy was presumed to exist, for among other reasons, because the theory of evolution by natural selection required it. Newer species could not be just newer, but had to be more advanced, more adaptively able than older species. And so, a sequence based on age also became a hierarchy of relative adaptive development, of evolutionary progress.

## A Modern Tree of Life

There is another tree of life that looks much the same as the one just described that is commonly conflated with it, though what it portrays is fundamentally different. This second tree of life is ahistorical—it does not attempt to lay out events that occurred at widely disparate points in time, but instead depicts a hierarchy for just one point in time, the present. It does not describe the progression of evolution, but its *outcome*—the hierarchical nature of life as we find it today. Thus, there is an evolutionary tree and a modern tree.

The difference between the two is not so much *where* species are placed, as *why*. For example, ancestors of today's bacteria are at the bottom of the evolutionary tree because they are thought to be among the earliest, and therefore least adaptively competent, species. But bacteria are also at the base of the modern tree, and this cannot be because of their early appearance, since the modern tree refers to the present time for all species. It shows *modern* bacteria. And there is no *a priori* reason for concluding that the adaptive poverty of bacteria in ancient times afflicts modern bacterial species. Quite to the contrary, bacteria are believed to have survived the longest, and survival is the key criterion of adaptive sufficiency. If anything, we would expect today's bacterial species to be especially well adapted.

Whether or not this is true, we certainly have no basis for concluding that they are the *least* advanced, or the least adaptively capable, species today. Then why are they at the bottom of the modern tree of life? Since the *modern* tree has nothing directly to do with historical age, their location must be determined by something else, some other criterion or factor.

Whatever that factor or criterion is, it must fulfill two objectives. First, it must allow for *arranging species in a sequence*, not of time but of some other sort, and second, this sequence must specify *a hierarchical order*. In the evolutionary tree, age serves as the basis for both—the first because the sequence of age is measured or otherwise understood, and the second because the theory of evolution requires a hierarchy based on age. For the modern tree, the first objective is fulfilled by and based on the *classification of species*. Organisms are classified, by which we mean that they are placed in different sets of species and groups of species by criteria that are usually rooted in anatomy. The relationship between these classified sets is then specified to indicate their closeness to or remoteness from each other. Done effectively, classification yields a sequence that specifies the varying degrees of relatedness of species.

But as with age, such a sequence has nothing to do with a hierarchy. It does not establish a *pecking order*. As with the evolutionary tree, that order is assumed. Also like it, it is not based on proof of a hierarchy, but on Darwinian principles that demand one. For the evolutionary tree we understand the hierarchy to be the *outcome* of natural selection, of evolution. For the modern tree, rather than being the *result* of natural selection, natural selection compels a hierarchy for its *future* action. Without a hierarchy of adaptive abilities, species would be on a par with each other, and with everything else being equal, there would be nothing for natural selection to select among. There could only be random occurrences, and as such, there would be no selective force. There would be, could be, no progress, no evolution.

There is another issue. Because historical age is vectorial—time only moves forward—we can at least imagine that age and adaptive competence are correlated. The sequence produced by classification on the other hand is not vectorial; it does not admit to direction and provides no basis for setting one. But a hierarchy needs direction; it can only occur in one way—up.

Three vectorial factors have played an informal role in fashioning the modern tree of life. I say that they have played an informal role, because they cannot be formally, logically justified. The first is evolutionary age itself. In this case, the sequence produced by time is simply mapped onto the age-independent tree, without any justification for doing so. The second concerns the progression of life from water to land. No doubt many of you have seen images in textbooks, magazines, and the like of life emerging from the sea onto land as part of the march of evolutionary progress. The tree of life usually reflects this event and others like it, even though we have no reason to presume that life lived in water is adaptively less advantaged than life lived on land.

Finally, there is complexity. It is our understanding that as evolution progressed, more complex species came into being. As such, those presumed to be more complex are positioned above those that appear less so. I say "presumed to be more complex" because as explained in the last chapter, we cannot specify an organism's complexity however it may seem to us. Hence a prescribed hierarchy of complexity just cannot be produced. Still, as a matter of common sense and ordinary perception, single-celled organisms are considered simpler than those with many cells, and likewise organisms that lack tissues and organs are thought to be simpler than those that sport them, and the tree's structure reflects such distinctions. Moreover, it should be kept in mind that a hierarchy of complexity, justified or not, is just that—a hierarchy of complexity. As explained, there is no basis for assuming a causal relationship between complexity and *adaptive competence*.

All things considered, we have no basis—at least, no justifiable basis—for producing an adaptive hierarchy of modern organisms based on these or any other criteria, even if we are confident that such a hierarchy exists because natural selection requires it. As explained, natural selection requires it because without a hierarchy of adaptive competence, there is no basis for selection. And so the difficulty remains. There is just no way to systematically order the modern hierarchy as a function of some specific factor or group of factors. This is not because we lack the knowledge or intellectual tools to do so, but because adaptive differences among organisms just cannot be systematized. This is because all living things on the planet today, whatever evolutionary path they have traveled, have evolved over the same duration. That is, they all presumably regress to first life, to the original ancestor of us all. As such, all species have had an equal opportunity, at least as a matter of time, to develop adaptations, leaving no *a priori* basis for assuming one to be more adaptively competent than another. Comparing bacteria to humans, we can no more place today's bacteria at the bottom of the modern tree than at the top, save analogy to their ancestors, nor put man at the top, rather than lower down, except for personal preference.

Thus, whatever evolution demands, differences in adaptive capacity are not systematic but are the consequence of the experiences of our ancestors. Each organism's, each modern species' adaptive competence derives from the unique, and to us unspecifiable, experiences of its ancestral line. It is these experiences, these sources of natural selection that have produced variability in adaptive competence. As a consequence, absent a systematic basis for producing a hierarchy, a modern tree of life just cannot be constructed; at least, it cannot be constructed rationally.

## *Progress, Smogress*

By the latter half of the twentieth century, trees of life, indeed the idea of biological progress itself, fell out of favor, as did the notion of human dominance, and I hope you can see why. For us, recent arrivals, *arrivistes*, a few million years old compared to the billions of years of bacterial life, we have not really been tested by time. Will disease, our despoliation of the planet, our weapons of mass destruction, or who knows what else, end human life in tens, hundreds, or thousands, not millions, much less billions, of years? For all our capabilities, we are a vulnerable and self-destructive species who might well disappear even more quickly than we arrived—a barely noticeable blip in life's long history.

Accordingly, only conceit makes us think that we are the

dominant, most well-adapted species on the planet today. Aren't the simple, ancient, ubiquitous, crafty, and persevering bacteria better able to survive adversity? In this light, isn't the view that man is responsible for despoliation of a planetary Garden of Eden somewhat solipsistic, given that bacteria (and viruses) kill other species with greater effect than humans? Is our impact on the planet greater than that of oxygen-producing plants or the great carbon dioxide-producing biomass of termites? It is said that we need to approach nature more humbly, more objectively, and when we do, we will see that the place of others is more imposing, significant, and secure, however smart and dexterous we are. Perhaps bacteria or cockroaches or who knows what else belong on the treetop, not us.

For reasons such as these, as well as the introduction of more accurate carbon dating methods and DNA analysis, the tree of life came to look more like a bush, with different shoots going off in different directions relatively early in life's history. Though some details of the old tree remained in place, after a common origin, major evolutionary paths seem to parallel each other, rather than being part of a single progressive series. Life seemed less a matter of hierarchy than of different organisms simply having followed different evolutionary paths. To try to specify which species today are more or less advanced, which are superior and which are inferior, has come to seem more like a moral than a scientific determination.

Nonetheless, as explained, notwithstanding our inability to impose intellectual order on differences in adaptive competence, they must exist if the theory of evolution by natural selection is correct. Some organisms, and by extension some species, must be better able to survive life's vicissitudes than others. Not only that, but as it happens many adaptive differences are pretty obvious, the stuff of everyday life. For example, a lame animal is less adaptively endowed than one that can effectively hunt or escape a predator, at least if hunting and escaping are crucial facts of its life. Likewise, and most generally, during development, a fertilized egg is less adaptively able than a newborn baby, and a human baby is less adaptively able than an adult human, and a human prepared for independent life by education, technical skills, strength, dexterity, perseverance, and on and on is more adaptively able than individuals who lack these abilities. Indeed, the differences in adaptive competence that we see in everyday life, much less in exotic situations, seem endless.

## Measuring Adaptive Capacity

Perhaps if we put aside notions of evolutionary progress and adaptive hierarchies, we can find redemption and understanding by discovering other ways to assess adaptive capacity. Perhaps the tools of materialist science can help us systematize, categorize, and evaluate differences in adaptive capacity. Indeed, the task, if not the measurements, seems pretty straightforward. All that must be done is determine a simple product—the product of the weighted average value of an organism's adaptations and their number. By "weighted average" I mean the value of a particular adaptation weighed against others. If this weighing can be done, and if we can also count the number of adaptations, the product would provide an objective "comprehensive adaptive index," a measure of the adaptive competence of an organism or species, without the need to impose some arbitrary hierarchy. With such numbers in hand, we would be able to quantitatively compare the adaptive competence of species.

Unfortunately, however straightforward, biologists can only produce such an index in theory, in their dreams. Both as a matter of principle and practice, no such accounting is possible, and no directory can be compiled. When all is said and done, establishing the adaptive capacity of an organism or species is a hopeless task and the reasons are imposing and instructive:

> ▶ First and foremost, there is no set of scientific principles that we can rely on to determine the weight to be given to any particular adaptation relative to any other. Beyond deciding whether an adaptation is both necessary and sufficient for life or merely sufficient, it seems that we have no

way—at least no objective way—of assessing their relative importance. And even if we are confident that one adaptation is more valuable than another, how can we assess how much more—two times, ten times, or a hundred times? For example, how would we go about weighing the adaptive value of vascular perfusion on which our life depends versus a membrane transporter in a bacterium that allows it to accumulate a nutrient on which its life depends? Should they be given equal weight because they are both critical for survival (necessary and sufficient), or because they serve similar functions (provide energy and matter to cells), or does vascular perfusion, requiring as it does the mechanistic and regulatory sophistication of the circulatory system, have a greater adaptive vibrancy than the bacterial transporter, which might only be a single protein?

▸ Second, since the worth of adaptations can only be known in action, they cannot be assessed *in advance of their actual use or deployment.* For adaptations that are not constantly in use, we can only know their effect after the events for which they have been marshaled come to pass. That is, we can only know their worth historically. Even if we have enough knowledge of similar events in the past to predict the likelihood of a particular outcome, we still cannot know the result of a specific event, or for that matter, what adaptive property (or properties)—claws, speed, strength, agility, or some combination—will turn out to be determinative in advance of a particular occurrence.

▸ Third, as with all things evolutionary, context is everything. What is a valuable adaptive trait in one environmental circumstance may be of no value or even a hindrance in another. For example, as noted, protective coloration is only "protective" if the organism sits on a background with which it blends. If the organism is on a substrate of contrasting color, it may be maladaptive, inviting attack. As such, its value can only be weighed in context. Some inherent worth cannot be ascribed to it.

▶ Fourth, we have no way of knowing whether any list of adaptive properties that we might compile is comprehensive. Indeed, it is probably fair to say that whatever our list, whatever the organism, even for the simplest, it would be incomplete. We would inevitably miss something, some action or circumstance of life. And even if our list were comprehensive, how would we ever know it? For complex species like us, it seems both numerically (a very large amount of information) and statistically (a very low probability that all would be included) close to, if not in fact, impossible to produce such a comprehensive list.

## *Sluggish Numbness*

Given these circumstances one might think that biologists would despair of ever determining adaptive competence, but they have not. They are wonderful weavers of life stories. For example, the quote from Walter Cannon in the epigraph at the beginning of this chapter tells us that amphibians such as frogs shut down their activities to survive cold temperatures. "They spend the winter in sluggish numbness." This is because frogs are cold-blooded (poikilotherms) and their body temperature mirrors that of the environment. When the temperature turns cold, their body temperature falls, and their level of activity is reduced accordingly. When it gets cold enough, frogs enter a torpid, but still living state.

On the other hand, with the exception of hibernation, warm-blooded animals (homeotherms)—humans and other mammals—maintain their body temperature when it gets cold. They may even increase their level of activity to maintain it (e.g., generate heat by shivering). Given this rather stark difference between amphibians and mammals, Cannon thought that mammals were better adapted than frogs. They retain all of their normal activities, all of their adaptive capabilities in the cold, whereas the lethargic amphibian does not. As the frog lies in the mud, it is vulnerable to environmental forces of all sorts, including attack by predators. On this basis, bolstered by the understanding that life on land came from and reflected an advance on sea-dwelling predecessors, it seemed reasonable to assume that the adaptively disadvantaged amphibian belongs lower down on the tree of life, and this is where they are commonly placed.

Following the same line of thought, we might ask how the frog's adaptation to cold compares to that of insects. As it happens temperatures at which frogs become dormant kill most insects. Might we not then say that amphibians are more advanced than insects, and that their dormant state is protective? Or how about comparing two homeotherms—say cows and humans?

A cow faced with cold weather will lie close to the ground to reduce its exposed surface area, as well as seek the protection of trees, bushes, or large rocks. At first glance, humans might seem to be at a disadvantage in this circumstance. Being smaller, we have a larger surface to volume ratio and consequently lose heat more readily than cows. But it is also true that by being smaller we can escape the cold and wind more easily and find a small niche to shield us, or even a cave to take us out of the weather. And of course, humans have many other ways of dealing with cold weather that are unavailable to cows and that mitigate our larger surface to volume ratio. We can build a fire to warm ourselves; put on gloves, a hat, and a coat (as my mother always exhorted); or build a protective structure. When all things are considered, humans seem far better adaptively endowed than cows to deal with cold exposure, and as such belong above them in our adaptive hierarchy.

But all we have done is evaluate adaptive capacity for *cold or winter weather*. How about hot weather, or a difference not due to weather? We can see the difficulty in trying to make such determinations with a silly example. It would not be surprising if a scientist who spent his or her life studying the nature of appendages on animals concluded that they are a critical adaptive feature (remember the "movers"). He or she might argue that animals with appendages are more advantaged than those lacking them, say comparing crabs to worms. In a fit of quantitative exultation, following this internal logic, and presuming that more is better, our scientist might conclude that species with eight legs are more adaptively capable than those with six (spiders are better adapted than insects), that six legs are better than four (insects are better adapted than vertebrates), and that the most well-adapted animal of all is the millipede with its many legs.

Of course, however enamored our scientist is with limbs, this would be a very foolish conclusion. He or she would not only have ignored the uncertain role of number in the effectiveness of locomotion, but all other adaptive features related to locomotion, such as the need to coordinate movement, speed, endurance, and dexterity. But even more importantly, he or she would have ignored every other adaptive property of life, period—all those *unrelated* to locomotion. Our scientist would have displayed a myopia born of a deep interest in appendages, the myopia of the "appendeologist."

The point is simply that we can imagine all sorts of adaptive hierarchies based on a particular attribute or group of attributes of our choosing. But any pecking order that we might create can be demolished as easily as it is contrived with another story. What if the torpid frog's color blended with the mud (as it does) and hid it from potential predators (as it does), and what if being in the mud protected it from the cold and wind (as it does), while our mammal, warm temperature and all, cannot find a place to hide from the weather or from predators? And what if the mammal cannot find food, while the frog doesn't require it? Now who is the most adaptively advantaged? Who belongs above whom in our adaptive hierarchy?

We could go on, adaptive activity by adaptive activity, creating and demolishing hierarchies at will by introducing additional variables. As said, why not a hierarchy for hot, rather than cold, weather? And why not one based on the thickness of an animal's fur, not its surface to volume ratio, or why not some combination of the two? Or how about hierarchies grounded in skill in obtaining food or protecting oneself from predators? Or how about how these properties are affected by the weather? Simply by changing or adding variables, we can obtain a different result, a different order that will demolish any hierarchy we previously fabricated. And of course, the choice of these variables is ours, not nature's, unlike the real world.

## Harmony

These are heartbreaking circumstances for science. If adaptations belong at life's center, as I believe they do, if they are the crux of what we mean when we say something is living, as I believe they are, then it turns out that what is central to life barely seems accessible to scientific evaluation. We cannot even apply a value, a number to their worth, or count them. Even in a controlled laboratory setting or on a circumscribed plot of land, characterizing organisms and environment as fully as we can, we are unable to assign a comprehensive adaptive index. The situation is either too restricted or too uncertain.

And this, though the literature in biology and medicine is filled with detailed information and precise measurements on the mechanisms that underlie biological adaptations. These mechanisms are found in and are about cells, the brain, the stomach and lungs, and on and on to include every tissue, organ, and system of each and every species. Indeed, it can be fairly said that our modern interest in the multitude of mechanisms that underlie biological processes is really about the mechanisms that lie beneath or give substance to biological adaptations. But for all its rigor and quantitation this knowledge does not help us assess their worth as *adaptations*. We can characterize adaptive qualities as being active, not passive; as either necessary and sufficient for life or just sufficient; facing inward or outward; fundamental, regulatory, or situational; responsive to continuous environmental forces or to periodic occurrences; responsive to constraints imposed by physical and mathematical law or the consequence of random or chance occurrences outside those laws. Yet we have no way of assigning a value to them, of assessing their worth, individually or together.

Nor can we predict the occurrence or result of future adaptive actions (save with some modest help from statistics), even though being able to accurately predict the future—like the trajectory of a tossed rock—is a central goal of scientific understanding. This is even true for such internal states as blood sugar levels. Though we can certainly measure them accurately, and take note of risk factors for diabetes, we cannot predict the future. And if the disease is contracted, we have no idea what its trajectory will be, what weakness will lead to death or when, or for that matter whether our demise will be due to diabetes at all, rather than some other cause, such as the weather or cancer. More broadly, though we can specify with great precision at what oxygen pressure vascular perfusion will cease being effective and death inevitable, we cannot predict when or why this will happen.

Due to such overwhelming uncertainties, despite our best efforts, we find ourselves leaving the world of science and being forced back into the world of subjective description. Invariably, and it seems, inescapably, ordering organisms in terms of their adaptive capacity, their ability to withstand one or another environmental circumstance, or such circumstances in general, a necessary predicate for their scientific evaluation, does not seem possible *despite the fact that we know that such differences, sometimes dramatic, exist between individuals and species.*

So although adaptations are life's essence, its basis, the outcome of the evolution of the species, and though we can describe the mechanisms that underlie these phenomena in their multitudes—species by species, organism by organism, circumstance by circumstance, often in great detail, and with substantial accuracy and precision—there seems to be no rigorous scientific instrument capable of grasping their worth or counting their number, much less applying such values to specific circumstances or future events. Science loses its power at this critical juncture. As the American comedian Jimmy Durante used to say, "Wadda purdicament!"

It may not seem like it, but *this* is the "harmony" of the chapter's title. As we seek life's essence, looking more and more deeply, we find the following:

- ▸ Rather than life becoming more and more remote from our everyday subjective experience—more and more the stuff of molecules and genes—the opposite is true.
- ▸ Life's deepest nature and the life we experience not only turn out to be connected, but the same. And finally,
- ▸ Neither materialist nor reductionist science appear able to serve as the vehicle for achieving this deeper knowledge of the phenomenon of life.

This is reminiscent of experimental psychology's situation during the twentieth century. As the field sought to understand the nature of mental processes by applying rigorous, objective, third-person scientific analysis, it found itself increasingly obliged to either relegate its central questions to the netherworld of epiphenomena, or label them of little interest and no consequence. As a result, the discipline's scientific reach was stunted and less relevant to its mission except perhaps as practical tools, even as the effort to understand expanded. The truly important questions were lumped together and put aside as something called "personality theory" defined by the *Encarta World English Dictionary* as—the sum of all "attitudes, interests, behavioral patterns, emotional responses, social roles, and other individual traits"—and their study was left to those unfortunates who clung stubbornly to an antiquated, subjective, and nonscientific point of view. Even today many neuroscientists digging deeper and deeper into less and less, into the details of receptors and synapses, are confident that they are on the road to understanding the wonders of the mind. They believe that this understanding will inevitably emerge from a darkened molecular tunnel in a blinding light of far-reaching enlightenment.

In the same way, the adaptive nature of life exposes the serious limitations of the biological materialist's approach to understanding what it means to say that some material object is alive. It and its reductionist underpinnings—its objective, third-person analysis—simply cannot explain life's deepest character. Though some still deny it, during the twentieth century physics came face-to-face with a similar problem. There was the uncertainty principle, quantum mechanics, relativity, string theory, multiverses, dimensions outside our comprehension, and beyond the big bang, where it all came from. As our knowledge of the physical world became deeper, against expectations, uncertainties seemed to grow, not diminish, and our understanding appeared to become increasingly abstract and obscure. Physics, it was said, seemed more like metaphysics than science.

For biology, the opposite has been true. The more we have come to understand the material nature of living things through genetics, chemistry, and structure, the more concrete and tangible life has seemed—as said, being neither more nor less than its molecules with their reactions and interactions taking place in space and over time. But as we have arrived at this understanding, it has become harder and harder to find life among its chemicals. Life itself has come to seem little more than a shadowy notion—an apparition, an ephemeral, fading epiphenomenon whose existence can only be affirmed by asserting it, and of course by our own very unscientific experience of it.

In this sense, *Life beyond Molecules and Genes* resurrects life. It declares that it is a real phenomenon found in the adaptive activities of living things, not their material underpinnings. Life, we have learned, at its deepest level looks exactly like life as it is experienced in the *real world*—uncertain, fragile, persevering, painful, cruel, perverse, joyful, and exultant. While it all emerges from material antecedents, life is not some sort of chemical or genetic abstraction. Whatever we would wish, the methods of modern analytic science that have served us so well otherwise leave us at sea in trying to obtain a truly deep understanding of the phenomenon we call life.

# Is There Life beyond the Genome?

—

As a narrative account of ordinary lives, genetic determinism
is hopelessly simple-minded.

**—JONATHAN FRANZEN, A NOVELIST, IN A LETTER TO THE**
**NEW YORK TIMES (7/29/03) ON DEPRESSION**

IN READING a recent series of essays in which the editor asked
"leading scientists" to predict the future of their field of interest
fifty years hence, I came across a most remarkable article, "Son of
Moore's Law," by science writer and evolutionary biologist Richard Dawkins (in *The Next Fifty Years,* ed. John Brockman [New
York: Vintage Books, 2002], 145–58). Dawkins' vision for biology fifty years down the road was awe-inspiring. He talked of
"three steps to 'computing' an animal from its genome":

▸ The first he said had been "completely solved." It is the
ability "to compute the amino acid sequence of a protein
from the nucleotide sequence of a gene." This is the genetic
code.

▸ The second is to compute the complete three-dimensional
structure of proteins from their sequence of amino acids.
This is not yet possible, and doing so is perhaps far in the
future, but still we know the detailed three-dimensional
structure of a rapidly growing list of proteins, numbering
in the hundreds, soon the thousands, by observing, not cal-
culating, molecular structure, primarily by imaging protein
crystals with X-rays.

▶ The third and, by Dawkins' estimation, the most difficult step, is to "compute the developing embryo from its
genes and their interaction with their environment," which
he says is comprised mostly of other genes. In this regard,
he expresses great hope for the Hox genes in providing a
genetic understanding of development.

Having laid out these goals, he then "conjectures" that with
this knowledge in hand fifty years hence, feeding the genome of
"an unknown animal into a computer" will yield nothing less
than a "full rendering of the adult animal." He imagines a detective taking blood from a crime scene and having the computer
issue forth the face of the suspect from his blood, even producing
a series of faces tracing the appearance of the individual "from
babyhood to dotage."

But Dawkins' prophecy does not end there. Quoting himself in
*Unweaving the Rainbow,* he says that "if we could only read the
language, the DNA of tuna and starfish would have 'sea' written into the text. The DNA of moles and earthworms would spell
'underground.'" And he goes on that by 2050, when we have
mastered the language of DNA, we will not only be able to produce a full rendering of an animal from its genome, but the whole
"detailed world in which its ancestors . . . lived, including their
predators and prey, parasites or hosts, nesting sites, and even
hopes and fears."

Has a more completely molecular description of life, including
as it does a complete historical accounting, ever been offered?
What historian wouldn't give his right eyetooth for such a "scientific" understanding of the past? Dawkins only makes these
predictions after providing us with the following caveat: "It has
been said often enough to become a platitude, but I had better
say it again: To know the genome of an animal is not the same
as to understand that animal." Okay, but why not? What is missing? What else is needed to understand the animal beyond its
genome? What is so familiar, so platitudinous, that it does not
need to be specified? What is missing, what is so platitudinous,
so familiar is nothing less than *life* itself.

## Is There Life beyond the Genome?

Not only is there life beyond the DNA genome, there is no life in the genome. As we have discussed, its necessity is not sufficient to produce life. Yet to the genetic determinist, this is exactly where life is to be found. Or to the more broadly thinking biological materialist, life is to be found in a particular complex physico-chemical system. And that system does not differ from other physico-chemical systems in any *fundamental* way, but merely in its specific embodiment. Viewed from this perspective, when you get right down to it, labeling one object living and another inanimate is really just a matter of convention or convenience. We could as well call a volcano living as plants for all the term means. They are nothing more than different physical and chemical systems.

This opinion is widely held, but it is far more than that. There is a vast body of experimental evidence accumulated over the past hundred years that supports it. Indeed, it is the unmistakable conclusion; I believe the inevitable conclusion of the kind of molecular reductionism that has dominated biological research in modern times. If life is physics and chemistry, then it follows simply and directly that a complete understanding of the phenomenon should be achievable *through* physics and chemistry—through the study of life's molecules, their reactions and interactions, especially those that involve DNA and proteins. Though it has only been made explicit from time to time, most recently, and with great fanfare by those promoting the Human Genome Project, this belief is implicit in an extremely broad body of molecular and biochemical research.

## The Search for the Holy Grail

Several years ago now, when the Genome Project—the Manhattan Project of biology—was considered complete and scientists had come close to deciphering the entire sequence of the immensely long DNA molecules that comprise our genetic endowment, it was heralded as a great scientific achievement. But if one read carefully, beneath the screaming headlines and hyperbolic press conferences, it did not live up to either its press clippings or its claims. Indeed, if one wanted to be a spoilsport, one could even say that for those who saw life as being manifest in our DNA, it was a colossal failure.

No comprehensive understanding of life dawned from the complete or almost complete uncovering of our genetic endowment. Unquestionably, it was a proud achievement, an enormous technical feat from which we have already begun to learn many useful things about DNA and our genetic constitution, and we will no doubt learn more for years to come. But it did not beget the millennial age of biology, as many seem to have expected.

More than that, the results of the Genome Project were ironic. Rather than forever installing molecular genetics as the key to understanding life, it brought this view under more critical scrutiny than it had experienced during the prior half century or so of genetic triumphalism. Sometime after the flood of positive press reports that followed publication of the genome, our local paper noted that the "Genome Discovery Shocks Scientists" (Tom Abate, *San Francisco Chronicle*, February 11, 2001). What was shocking was that species like the lowly roundworm, comprised of several hundred cells, lacking all the wonders of the physiology of complex species such as man, with our trillion or so cells, had almost as many genes, not many orders of magnitude fewer as expected.

In a sense this was not surprising. As discussed above, in spite of life's many variations, cells rely on similar protein products derived from similar DNA sequences to carry out their basic activities. They share a common chemistry. But in another way, it was shocking. As modern genetics understands it, inherent to DNA's codes for proteins are life's "potentialities." Through the intermediary of protein action, they give rise to each and every feature, or at least each and every major feature of living things. We understand these various codes to be the "genes" that Mendel and the classical geneticists described, the elements (Mendel's and Aristotle's term) that underlie life's observable features, the genotype to life's phenotype.

Against all expectations, the Genome Project taught us that this is not the case. And albeit negative, this is its true accomplishment. As explained, we have learned that most nonbacterial cells—whether from plants, roundworms, humans, or anything else thus far examined—have roughly the same number of genes, and that the small differences in gene number that are found do not seem likely to account for the enormous differences in the way we look, or how we are organized and function.

Yet however great this falsification, DNA science was not deterred. The disappointment was quickly set aside, victory proclaimed, and the course for the future charted. With what seemed barely a backward glance or contemplative pause, the problem, it was decided, was not in looking for life in DNA, or in the materialist perspective more generally, but as Dawkins explains, in the fact that we simply did not understand DNA and its protein products well enough, at least not yet.

Counting genes on DNA, just months before, the profound undertaking that was going to expose all was now seen as "simplistic," as many claimed they knew all along. What science needed were more resources devoted to better understand, better explicate the molecular basis of life's diversity. Some thought that the Promised Land could be reached if we could just provide a complete accounting of protein molecules, in line with the second of Dawkins' three prongs.

Others thought his third requirement was key. To obtain complete understanding, we had to master the nature and character of the embryological development of organisms, particularly the *regulation of gene expression*. As Dawkins assures us, when all three prongs have been realized, the journey from Mendel's peas to the full comprehension of life's nature in genetic terms will be complete. Admittedly a tough, perhaps the toughest, part of the passage from innocence to maturity lay ahead. It might take a long time, but when we finally understand genomics (DNA), proteomics (proteins), and the regulation of gene expression (the new genetic "regulatory biology" and its close relative developmental biology [or "evodevo"]) in all necessary detail, we will be able to explain the phenotypic differences between species and spell life out in wholly molecular terms.

I believe that this hope is a fantasy, like Dawkins' admitted fantasy, or it simply misses the point. There is no possibility, no possibility at all, that we can explain *why* hearts beat or birds fly in terms of molecules alone, even in terms of those most remarkable DNA and protein molecules, even if we knew all there was to know about them. However imagined, however understood, they cannot account for the wonders of the bird's song, the birth of a new baby, or the achievements of the human mind, among a myriad of other things. Deny as we may, it is an unavoidable fact of nature that the phenomenon of life transcends its molecular antecedents.

A biologist who thinks that all can be explained in terms of molecules is not much different than a physicist who thinks that a complete understanding of *living* beings can be achieved by comprehensive and detailed knowledge of the *atoms* of which we are comprised. After all, we are comprised of atoms, are we not? Most scientists today, even physicists, would agree that such a claim is silly (but see Mark Buchanan, *The Social Atom* [New York: Bloomsbury USA, 2007]). With such knowledge alone, however detailed or comprehensive, we would not even know about life's chemistry and genetics, no less about life itself.

## *Adaptations as Life*

However critical, however illuminating, however necessary, life cannot be understood in terms of its DNA or its materiality more generally, not in its particular parts nor taken as a whole. As I have argued here, to attempt to do so is to squeeze life out of the living, to desiccate it. This predilection for chemical explanation (including modern molecular genetics), for desiccation, by necessity ignores or elides the sufficient causes of the phenomenon of life.

In this rendering, features thought necessary for life are assumed (usually implicitly and without reasoned exegesis) to also be sufficient for its expression. Certainly, without chemistry, without DNA and proteins, without enzymes, without chemical synthesis, degradation, and transformation, without chemical energy, there can be no life. But as I have submitted, molecules, reactions, even their regulation, interconnections, and the complex consequences of it all, do not in and of themselves, singly or together, make an object alive. To say that they do is to mistake life's material features—its substance and organization—for being alive.

In the glow of reductionism's glorious successes in uncovering life's material nature, many biologists have forgotten, or perhaps never learned, how to think about living things in terms that transcend their molecular, chemical, physical, structural, and anatomical antecedents. They may know a great deal about life's *necessary* physical and chemical properties, but its *sufficient* causes.

But to disregard them is to disregard life. Only in their presence can a body be said to be alive. And as explained, we find life's sufficient causes in the adaptive properties of organisms. As Darwin and Wallace told us, adaptations bestow fitness in our struggle for survival, but they do more than that—they make us alive. However counterintuitive, adaptations are not features of otherwise living things, they are life itself.

## Ah, Sweet Mystery of Life

And so, unless and until we look beyond life's necessary features, beyond our modern infatuation with molecules and our predilection for molecular explanations, we will continue to believe against reality that they are by definition synonymous with life's sufficient causes. We will simply leap blindfolded from the world of necessity into that of sufficiency, bluffing or deluding ourselves into thinking that they are one in the same thing. Or barring such an indefensible act, we may hope against hope that someday we will be able to build a bridge of molecular and genetic understanding across the abyss, the hollow place, from life's molecules and genes to the amazing phenomenon of life.

As the Genome Project attests, modern experimental biology has expressed and acted on the faith that to understand what we are made of, to understand our parts, our substructure, our molecular substance, is to understand life, to understand what we are. It has been argued that it is a foolish waste of precious time and money to focus on life's *superficial* aspects, its adaptations to the environment. However fascinating such phenomena may be, when you get right down to it, studying them is just a form of stamp collecting—a hobby, a distraction from gaining a deeper scientific grasp of the underlying chemical and physical basis of life. And if we hope to obtain that knowledge, we must not be distracted.

But by focusing solely on events at the molecular (or for that matter, cellular) level, and viewing the properties of whole organisms, embedded in their varied environments, as a superficiality of no great scientific interest, modern molecular and genetic biology has made itself blind to life's true nature. Despite our desire for enlightenment, we have become sleepwalkers, feeling our way in the dark—molecular detail after molecular detail—hoping against hope to find the subtleties and mysteries of life by drawing inferences about life's deepest nature from modern chemistry and genetics.

# The Religious Allusions of
## *Life beyond Molecules and Genes*

THOUGH NOT A work of philosophy, and despite my initial pro-testations that it is a book about biology, *Life beyond Molecules and Genes* is undoubtedly about metaphysics if we define meta-physics as "the study of the nature of being and beings, of exis-tence and causality," of what comes after or lies beyond "physics," beyond the material nature of things. Indeed, it is fair to say that is what it is all about. On the other hand, its religious impli-cations—not religious observance or adherence, but numinous claims about reality centered on a divine involvement in the uni-verse and life—are more difficult to apprehend.

To begin with, though science is the product of religion, its mandate is to explain nature *without* invoking the deity. It does not even allow a pantheistic God into its embrace. Its purpose is to enlighten us about nature solely through the use of the human faculties of observation, reason, and logic, absent faith and tran-scendent belief. As such, it excludes God as an agent of causa-tion *a priori*. This does not mean that it denies God's existence or causative agency, or that by this disposition scientists are athe-ists. Many of history's greatest scientists were clergymen or great believers otherwise. But whether a scientist is devout or an athe-ist, science blinds itself to questions of faith and belief by inten-tion. They are simply not instruments of science.

As a consequence, one might conclude that science has noth-ing much to say about the involvement of the Divine in life, either as we live it or as it came to be, and that as such the opinions of scientists about religion are simply matters of personal persuasion that, however thoughtful, cannot be informed by science. And yet, it has been common in modern times for scientists and laypersons alike to point to scientific ideas and the results of scientific research to *debunk* the presence of God. But how can this be? How can science at one and the same time absent itself from a consideration of God and use its results and theories to assess the belief? Confidence that it can do just that is perhaps science's greatest conceit.

## Science Debunking God

In a modern context and in relation to life, the argument goes something like this: the propositions that "life is God's creation," "atoms underlie matter," and "genes underlie life's form and function" are comparable notions that propose abstract causes for properties of nature. As such, they should be correspondingly amenable to the ministrations of science. Science can seek evidence of God, just as it can of atoms and genes.

But is this true? Are these truly parallel ideas? Is the concept of God really like that of atoms and genes? Since atoms are proposed to underlie all matter and genes all life, we have looked for and found them in the material objects thought to contain them. But how would we look for God? If he[1] were of the universe, beyond the pantheists' belief that God is one and the same as the material world, how would we identify him immanent to and yet different and distinct from its ordinary material contents? What distinguishing feature or features would we look for? What are God's *properties*? We might say that he is eternal, neither created nor destroyed, and look for him in what appears eternal in the material world. This would, of course, exclude living things that are constantly being created and destroyed, but

---

1. For those who may find it offensive, the use of the male personal pronoun for God is merely a traditional convenience. God is not only meant to be gender neutral, but independent of gender.

perhaps God can be found in some concurrence with the sub-stituent atoms of matter, the particles that form them, or energy itself. However, other than claiming an identity, how would we go about identifying God juxtaposed and yet set apart from these material characteristics?

But whether or not God can be found in our world, as its cre-ator and demiurge, he must also transcend it, be outside it. And so, we must look for his presence elsewhere as well in the mate-rial world. But how can we do this? How can we look outside the universe or universes? Where would that somewhere else be? Would it be anywhere? Even if we could figure out how to do this, as with God's immanence, what would we look for? God may be all-powerful, all-knowing, for some loving and benevo-lent, for others judgmental and punitive, and of course, the cre-ator of all we see before us, but of what use is such knowledge to science? How can we look for these as properties of God? In any event, and more generally, what tools do we have to identify his transcendent presence?

The answer is none, at least if we are talking about scientific tools. The simple fact is that we do not even know what questions to ask, much less that any that we might construct would make sense in God's hidden world. Not only don't we know God's prop-erties and means, or how to find out about them, outside of reli-gious belief we have no idea of his wishes or desires, his thought processes or motives, and for that matter, even if he has them.

After all is said and done, the only thing that we can do is look for signs of God by imagining what kind of world he would cre-ate and how he would go about creating it, and then comparing these imaginings to the real world we see before us. Unlike believ-ers, science cannot simply say that whatever its attributes, this is his world, the world he created, and he created it as he saw fit. It must wonder what God would *want* to build and how he would *want* to build it. But inescapably such speculations come at an enormous price—that price being an unavoidable and immense presumption—what would I do, if I were God?

Perhaps the arrogance of such a question should dissuade us

from asking it. But this has not been the case, and some, ironically including atheists, presume knowledge of God's aims and desires, and then armed with their conjectures, look for his presence. There have been two approaches. We can call the first epistemological because it examines the text of the Hebrew Bible, though in point of fact it is antiepistemological, since it refuses to analyze the text, but instead insists on taking it at its word, taking what it says literally. Having done this, it then assesses what the Bible has to say in light of modern scientific knowledge. Most famously, science has criticized the biblical account of the origin and evolution of life given in Genesis.

Fossil evidence has made it clear that biological evolution occurred over billions of years, rather than life being planted in all its variety and abundance on the planet in the six days allotted by Genesis. As such, from the standpoint of science, the biblical story of life's origin is incorrect. For some this means that science has proven the great book, for all its wisdom and beauty, no more than a document written by mortals ignorant of modern science, and certainly not the product of a Divine presence. Why would God sanction such a fairy tale?

But others, including the likes of Philo, St. Augustine, and Maimonides, wondered whether a literal reading of Genesis and the Bible more generally makes any sense. For example, wouldn't an all-knowing and infinitely talented God have had the Bible's writers create a glorious poetic allegory for our superstitious ancestors as well as our ignorant selves? Indeed, even our scientific knowledge, so carefully and thoughtfully accumulated, might be little more than an allegory provided by God. Anyway, and more mundanely, what does Genesis mean by a "day"? As the psalmist says, "For a thousand years in thy sight are but as yesterday when it is past" (90:4). Whatever the Bible intends, it is one thing for science to question the literal accuracy of biblical descriptions, and quite another to use this analysis to argue that God is a fallacy.

The other approach is ontological and twofold because it concerns being —the state that is life, and becoming—its evolution.

It is thought by many to offer the most consequential *scientific* arguments against God. Let us take "becoming" first. The argument has been made that if becoming, that is, if evolution is the result of natural selection—as most scientists today believe and as a significant body of evidence suggests—then the world is godless and its forces disinterested in life. From this point of view, evolution and its mechanism—natural selection—is a haphazard business, the result of chance occurrences that care not a whit about our personal survival or, for that matter, the survival of life. Unless godly design can somehow be discovered amidst such circumstances, God neither conceived nor fabricated life's evolution.

Yet as *Life beyond Molecules and Genes* explains, however uncaring natural selection may be, it is not haphazard. While adventitious events are critical, life's central features evolved in accordance with the abiding laws of nature. They, more than anything, have made us what we are. Consequently, if natural selection is the mechanism of evolution, then it is not just the result of accidental, undirected events, but of a great and unbending fidelity to the laws of the physical world, the laws of mathematics, geometry, diffusion, thermodynamics, classical and quantum mechanics, and, of course, the laws of chemistry. Living things, most importantly, their adaptations, hew to them all. They shape life and craft its multitudinous forms and functions. And so, the claim that we can exclude God's hand from evolution because of its haphazard nature is false. Evolution has not been haphazard. And even when events are haphazard, outcomes are not; they invariably cleave to nature's laws. To use a term often larded with implications of the Divine, the laws that govern the physical world have *designed* life, whether or not they are God's laws.

The other argument against God has to do with "being," and comes from the understanding that life is simply a particular physico-chemical system. Life, it is said, is found in our substance, in our DNA and proteins, in the godless world of chemistry and physics, and as such, has everything to do with matter, and absolutely nothing to do with God, except as human superstition. This view is bolstered by the remarkable power of scientific measurements—their precision, clarity about what is being measured, and reproducibility. They assure us that our conclusions are rigorous and, as such, seemingly indisputable. They allow us to say that the results of scientific research, the measurements we make, are *objective*, signifying that they have meaning independent of the thoughts that give rise to them, independent of our reasons for making them. As such, they uncover the objective reality of the material world, a reality in which there is neither need for nor room for God. But as *Life beyond Molecules and Genes* shows, however authoritative the measurement, the belief that life can be found in mere matter is badly mistaken. Life is not a particular material embodiment. It is not intrinsic to certain chemicals and their reactions, to certain anatomical structures and physical states, but is found, and found only, in our adaptations, in what we do and why we do it. As such, life transcends its material incarnation. It not only admits to spirit, it is spirit. It not only allows transcendence, it is transcendent. And it not only can accommodate contingent causes, it is contingent. Though based on chemistry and physics and, of course, our all-important genes, life is an immaterial marvel.

Consider the example of a conversation. Two material entities must of course exist for a conversation between them to take place. But is it therefore correct to say, as we often do, that the conversation takes place between these two objects, between two individuals? Of course bodies, even human bodies, do not converse *per se*. They might be dead, sleeping, silent, disinterested, resistant, or like babies or the deaf before learning how to communicate, unable to converse. And certainly we cannot say that the muscles that allow speech, such as those of the larynx, pharynx, and mouth, converse. Nor can conversation be found in the sound waves they emit, or in the fact that we can hear each other's utterances. Indeed, a conversation might not involve sound waves at all, but be carried out with gestures or in writing. Naturally, conversation is found in what the two parties have to say, and what they have to say is not a matter of matter, of mechanics, of sound waves, or even the electrical activity of the brain. It takes place, and only takes place when one living being shares certain immaterial, transcendent, and contingent properties called thoughts with another.

And so we cannot claim that because life is a material phenomenon that God's presence can be excluded, because life is *not* found in mere matter, but in properties without substance that reside and flourish in a virtual space, similar to, some might say exactly like, the one that religion associates with the soul. The modern materialist will point out that like all of life's immaterial aspects, conversation arises from its material necessities, from our brain cells, mouth, and larynx, from sound waves, auditory receptors, and whatever other physical means we use to communicate. And that as such, conversation is part of the same material world as everything else. In this view, life's material nature can only be fully understood by including and embracing its immaterial products, and we need not conjure the Divine to do so.

Yet, isn't this just a rationalization to deny what is self-evidently a nonmaterial event? If everything is part of the material world, then it should come as no shock that nothing exists outside it, including God. But then aren't we simply *defining* God out of existence? Even more problematic, if our claims that God can be dismissed because evolution is a haphazard affair and that life has no room for God because it is just a material matter are *false*, then aren't we as scientists obliged to conclude that the presence of design and an immaterial life are proof of God's dominion? If our suppositions about God's ways are correct, then haven't we proven his presence?

If we say, as I believe science must, that this is not necessarily so, that these may not be God's ways after all, then the whole enterprise of looking for God in the character of the material world and beyond fails. How, as a matter of logic and science, can we on the one hand say that God would do things this way, and yet when we find evidence that he does, turn around and say that this may not be his approach after all, and that what we see may be due to some other agency? We can do this because in the end, we cannot avoid the fact that God's ways and objectives are unknown to us whatever we might imagine.

Perhaps we can fare better if we try the opposite approach: if we try to debunk *science* with God. An evocative and fascinating claim made by Intelligent Design advocates is that natural selection cannot have given rise to the complex mechanisms that underlie so many biological processes, because, according to science, they would have to be accreted bit by bit, one protein molecule at a time. This is a problem because the theory of evolution says that things only evolve if they first have adaptive value, and adaptive value only inures when mechanisms are complete and functional, when all the proteins are in place. How then could the complex mechanisms of life have evolved piecemeal and in the absence of adaptive purpose if the theory of evolution by natural selection is correct?

This is a strong argument for life having been intended, planned by some extrinsic agency, force, or being, perhaps a Divine presence, rather than being the product of the impartial influence of natural selection. And yet, who can mandate that partial mechanisms had *no* adaptive value, perhaps of a different or less developed kind? But more importantly, whether they did or did not, this exposes the general difficulty in trying to determine God's presence or absence as a matter of science.

If for the sake of discussion we imagine that there is convincing proof that natural selection is *not* the mechanism of evolution, where would that leave us? What affirmative scientific conclusion could we draw from this knowledge? Most significantly, as a matter of science, would we now be obliged to conclude that biological evolution is of God's design?

The answer is not only no, but science cannot under any circumstance accede to this conclusion. The simple unadulterated fact is that doing so would spell its end. If every time a scientific idea was found wanting, we had the option of explaining the mystery by saying that God did it, science would disappear in a wave of godly accounts. That science cannot allow such a conclusion is an absolute requirement for its existence, as well as in a sense its *raison d'être*. As explained, science was only able to come into being as a mature discipline when it excluded God and matters of faith and religious belief from its method.

From science's standpoint, rejecting natural selection without any credible alternative scientific explanation for evolution would simply mean, and could only mean, that science is ignorant, that our current scientific concepts are inadequate, not that life is God's creation. Even if by some nonscientific means we were to find that life is attributable to God, science, *in its own terms*, would nonetheless have to reject the conclusion. This would not make it false. It just would not be a matter for science. Alas, allowing God and faith into science destroys it.

## Science as Religion

After all is said and done, we cannot say what evidence for or against God's authority would look like. It may make sense to us that the absence of design means the absence of God, or that the absence of natural selection means the presence of God, but we may be wrong on both accounts. The absence of what we think of as design may simply mean that his design is beyond our grasp, and the absence of natural selection may simply mean that another mechanism, presently unknown to us, is responsible that does not require or even allow God. Quite simply, science is in no position to make such determinations, to set rules to assess God's desires or to state what evidence for or against God would look like, what properties would have to exist or could not exist. Quite simply, God is beyond science, and hence the need for faith. As suggested, if scientists could set the rules, if they could make such determinations, they would be gods themselves.

So those, like atheists and Marxists, who think that the question of God's existence can be subjected to scientific test, are mistaken. Science can neither affirm nor disprove his existence without imposing its own religion. In a recent article, the philosopher David Berlinski warns scientists against serving as "God of the Gaps"—of inserting themselves in God's place to explain what we do not understand (*Commentary* magazine, April 2008). Humans as well as other species fill gaps in their understanding with their best guesses, with hunches, with what can only be unsubstantiated beliefs. We have little choice in the matter. Without being able to guess about uncertain circumstances, we would be immobilized in face of them and could not act. It is much the same for science. Science has an abundance of large and small gaps of understanding. For biology, consider the following impressive list:

- ► How exactly did life arise from its inanimate precursors?
- ► What is the mechanism of speciation?
- ► How did natural selection produce the immense complexity that is life?
- ► How do genes give rise to life (what is the relationship between genotype and phenotype)?
- ► To what are the huge phenotypic differences between humans and monkeys due? And finally,
- ► How does mind arise from brain?

In the normal course of events, scientists fill each of these gaps, and many lesser ones, with suppositions for which they lack proof. Nature abhors a vacuum, and so do scientists. Just as in ordinary life, scientists need a whole story. They need it to test their ideas in the laboratory, to teach the subject to others, and simply to assuage their uncertainty and insecurity so that they have the courage to continue. The problem is that in making their stories whole, they often come to believe what they have fabricated, they come to trust that what is supposed reflects what is true, and that they will eventually be able to prove it.

But such confidence is not merely unearned, it is an abomination. First scientists exclude God, and then assume his divinity for themselves. Divinity, because in fact they have no idea if the gap will ever be filled, much less in line with their notions, nor even that in filling it previously unrealized difficulties will not emerge that will lead to no less than the total collapse of their worldview. With a wave of the hand, scientists dispatch God and make themselves all-knowing master conjurers, Nostradamus in denial.

Furthermore, unlike religion, the arc of science is always temporal. All we have is our present understanding of things with its manifold deficiencies. Some say that however confident we may be that today's science provides an accurate account of nature's properties, that account will in one way or another be found lacking if not totally false in the future. However certain things may seem today, science is eternally tentative.

In any event, like it or not, scientists are not gods; we may be know-it-alls, but we are not all-knowing. It should be plain even to the most egotistical among us that humans do not have the capacity for limitless understanding. Despite our best efforts there are many things about the universe and about life that we do not understand, and no doubt there are many more that we are *incapable* of understanding or even imagining. How arrogant is the claim that we puny humans, mere specks in the universe, even if "made in God's image," are capable of grasping everything there is to grasp?

While our brain is a remarkable and inordinately complex structure, with more cells than galaxies, more connections than atoms in the universe, and we are doubtless capable of remarkable thoughts well beyond that of all other living things (at least on this planet), and that the results of our efforts can be immensely enriching, even noble, the brain is nonetheless a finite, rather smallish piece of ordinary matter, and if it is indeed the result of natural selection, then it is nothing more than a makeshift product of its labors.

As scientists we are told to walk humbly before nature, to be aware of our impressive limitations in face of its awesome character. One of these limitations is that science has no way of proving or disproving matters of faith or belief, such as God's existence. The fact is that both our most far-reaching and most personal questions are imponderable to science. They have been, and perhaps always will be the province of other kinds of thought— centrally, religious thought. At the end of the day, as powerful as science is, it cannot relieve us of ultimate questions of faith and belief, of God and his communion. Such questions, alike and apart from those of science, are part of the human condition.

.

# Notes and Suggested Reading

THE REFERENCES listed below do not begin to provide a comprehensive account of the ideas or writing on the subject of life's nature. It is a selection of volumes, mostly modern, from the mid-nineteenth century forward, most, but not all, of which are from a scientific or biological perspective, including technical material, textbooks, and writing for a general audience. Some citations are included for their relevance to a particular point or subject, others simply as a place to begin further reading about what is an extraordinarily broad topic. Where possible I have tried to keep the selections recent. With some important exceptions, and for better or worse, they are my personal choices as exemplars of a particular viewpoint, often chosen from among various reasonable possibilities.

Preface. As a narrative for a particular explanation of what makes certain material objects alive, *Life beyond Molecules and Genes* is not intended to summarize, much less compare and contrast, all such points of view, historical or modern. Doing so would no doubt be a valuable undertaking, but it would be an enormous task, and well beyond this book's scope. Not only that, but it would most surely distract from and might even obscure *Life beyond Molecules and Genes*'s straightforward argument and line of reasoning in a mélange of disparate and far from equally justified views.

Nonetheless, two comparisons simply cannot be avoided. This is because it is against them that the ideas offered here are proffered and should be judged. The first concerns the view—common

among both the general public and the scientific community—
that life is inherent to its material embodiment, what I have called
"biological materialism." The second is the idea that life can be
found in the complex nature of that embodiment, the subject of
chapter 10, "Life as Complexity."

Underlying *Life beyond Molecules and Genes*'s argument is
the claim—in accord with reason, Ockham's razor and thought
on the subject for millennia—that however varied, life is a single phenomenon. As such, it does not admit to multiple, dissimilar definitions. It cannot at the same time be this and that. It
is one particular thing and it alone. This said, it is frequently
claimed that it can be different things to different people, different things to different disciplines, and of such defined in different,
equally acceptable ways. To give two examples, in *On the Nature
and Origin of Life* (New York: McGraw-Hill, 1971), the philosopher Hilde S. Hein opens her discussion of life's nature by telling us that life admits to a variety of definitions depending upon
one's scientific or philosophical perspective. Yet she does not tell
us why this is so, or how it accords with logic or nature. The
famous geneticist and evolutionary biologist C. H. Waddington
envisioned two definitions of life in his *The Nature of Life* (New
York: Atheneum, 1962). Life, he said, could either be understood
from an atomistic perspective—from the point of view of molecules—or in terms of a continuum, meaning everything about
life that is not easily reducible to molecules. But having said this,
he does not go on to enlighten us as to how viewing it either way
explains aliveness. Attempts to define life in different ways that
are thought to be equally fitting sometimes seem more a reflection of the desire to accommodate different viewpoints than a
search for the truth. Or it may be a veiled acknowledgment that
that author is unable to provide a clear definition of the phenomenon. It may be this, but then again, it may be that.

Introduction. Most writing today in and about biology and medicine—research articles, technical monographs, textbooks, as well
as books for a general audience—imply that life and its chemical
and physical characteristics are one and the same thing. Life is

understood to be a particular physico-chemical or "mechanical" system, and nothing more. This view has its roots in the early seventeenth-century ideas of René Descartes. Life, Descartes said, is a machine, except for the transcendent nature of the human mind and spirit (*Philosophical Works*, vols. 1 and 2, trans. Elizabeth Haldane and G. R. T. Ross [1628; repr., Cambridge: Cambridge University Press, 1911]). Descartes' mind/body dualism was eventually replaced with a fully mechanical account of living things, including body *and* mind. See Julien Le Mettrie, *Man the Machine* (1748; repr., LaSalle, IL: Open Court Publishing, 1961).

An influential expression of this viewpoint, and a foundational and prescient document of the molecular biology movement, is the famous physicist Erwin Schrödinger's *What Is Life?* (Cambridge: Cambridge University Press, 1944). For a similar, more recent perspective, see the writings of Nobel Prize winner Francois Jacob (*The Logic of Life* [New York: Pantheon Books, 1970], and *The Actual and the Possible* [New York: Pantheon Books, 1982]). The most extreme form of this view is known as genetic determinism, the notion that life and its genes (DNA) are for all intents and purposes congruent. Though few are willing to consider themselves holders of "extreme" views, genetic determinism can be discovered, if only implicitly, in any number of books for a general audience, in many different kinds of biology textbooks, as well as in innumerable technical and general articles in journals, magazines, newspapers, etc., particularly those published in the years just prior to and in the wake of the sequencing of the human genome. For example, see science writer Matt Ridley's *Genome* (New York: HarperCollins, 1999). This point of view has been most vividly and effectively argued in a series of popular books by evolutionary biologist Richard Dawkins—for example, *The Selfish Gene* (Oxford: Oxford University Press, 1976), *The Blind Watchmaker* (New York: W. W. Norton, 1986), *Climbing Mount Improbable* (New York: W. W. Norton, 1996), and most recently, *The God Delusion* (London: Bantam Press, 2006).

The view that life transcends its material embodiment is found mainly in the writings of philosophers, with a smattering of volumes by biologists. The problem with most, if not all, of these often-thoughtful efforts is that in the end, they tell us little that is concrete about life. Even in modern times, the great gulf between the authors' ideas about life and the real world of biology, with its multitude of physical incarnations, cannot be bridged, and much like thinkers of old, they leave us with mere words to describe vague and unknown things. Though some make convincing arguments that life's material embodiment is not sufficient to explain aliveness, they invariably fail to offer a substantive affirmative account of life's cause. Either they propose nothing, something nebulous, or something mysterious. As a consequence, to compare these ideas to each other or to the view embodied in *Life beyond Molecules and Genes* is, for other than historical reasons, pointless.

This said, and his aversion to Darwin's theory aside, Henri Bergson's *Creative Evolution,* trans. Arthur Mitchell (New York: Henry Holt, 1911), written at the end of the nineteenth century without access to modern scientific understanding is a prophetic and discerning book. Also, with access to modern science, there is Hans Jonas' insightful *The Phenomenon of Life* (New York: Harper and Row, 1966). A list of other modern writings about life's nature beyond its material incarnation includes Hans Driesch's controversial notion of Entelechy in *Mind and Body,* trans. Theodore Besterman (London: Methuen, 1927), and *The Science and Philosophy of the Organism,* 2 vols. (London: Adam and Charles Black, 1909); A. N. Whitehead's *The Concept of Nature* (Cambridge: Cambridge University Press, 1920); Helmuth Plessner's *Die Stuffen des Organischen und der Mensch* (Berlin: Walter de Gruyter, 1975) and *The Limits of Community,* trans. Andrew Wallace (Amherst, NY: Humanity Books, 1999); Adolph Portmann in *Animals as Social Beings* (New York: Viking Press, 1961); Marjorie Grene's *Approaches to a Philosophical Biology* (New York: Basic Books, 1965); C. D. Darlington in *The Evolution of Man and Society* (New York: Simon and Schuster, 1969); Karl Popper's and John C. Eccles' *The Self and Its Brain* (Berlin: Springer-Verlag, 1977); Theodosius Dobzhansky's and

Ernest Boesiger's *Human Culture* (New York: Columbia University Press, 1983); Walter M. Elsasser in *Reflections on a Theory of Organisms* (Frelighsburg, Quebec: Orbis, 1987); Lynn Margulis' and Dorion Sagan's *What Is Life?* (New York: Simon and Schuster, 1995); M. Boden's *The Philosophy of Artificial Life* (particularly the articles by Ray, Langton, and Bedau) (Oxford: Oxford University Press, 1996); and Steven Rose's *Lifelines* (Oxford: Oxford University Press, 1997). Also see the short essay by Stuart Kauffman, "What Is Life?" in *The Next Fifty Years,* ed. John Brockman (New York: Vintage Books, 2002). Michael Ruse provides brief readings on the subject from the viewpoints of a variety of philosophers and scientists in *Philosophy of Biology,* 2nd ed. (Amherst, NY: Prometheus Books, 2007). Additionally, his and David L. Hull's *The Philosophy of Biology* (New York: Oxford University Press, 1998) and *The Cambridge Companion to the Philosophy of Biology* (Cambridge: Cambridge University Press, 2007) provide additional essays on the subject. For the view that life is to be found in its complex nature, see the notes to chapter 10 below. For a systems theory or relational approach, see Robert Rosen's *Life Itself* (New York: Columbia University Press, 1991). For yet other ideas, see Wendell Berry's *Life Is a Miracle* (Washington, D.C.: Counterpoint, 2001); Edward O. Wilson's *Consilience* (New York: Knopf, 1998); and finally, for an engaging introduction to Buddhist thought on life's nature, especially its doctrine of emptiness, see the recent book by the Dalai Lama, *The Universe in a Single Atom* (New York: Broadway Books, 2005).

The most radical questioning of a mechanical view of life has not unexpectedly come from religion. The complaint dates back hundreds, if not thousands, of years as various religious thinkers tried to join nature and religion. This effort, often subsumed under the rubric of *natural theology,* has and continues to play an important role in the conflict between various religious beliefs and science about the evolution of life. See the seminal *Natural Theology* by William Paley (1802; repr., Oxford University Press, New York, 2006). (A prior occupant of Darwin's

Cambridge rooms, Paley's "ghost" and religious/naturalist ideas about life's nature challenged the young Darwin to seek other explanations.) Also see the famous physiologist Charles Bell's *The Hand. Its Mechanism and Vital Endowments as Evincing Design* (London: William Pickering, 1837); Arthur O. Lovejoy's *The Great Chain of Being* (Cambridge, MA: Harvard University Press, 1936); E. S. Russell's *The Directiveness of Organic Activities* (Cambridge: Cambridge University Press, 1945); and Alan Olding's *Modern Biology and Natural Theology* (London: Routledge, 1991). The current variant of natural theology and the subject of much heated discussion today is known as Intelligent Design. It focuses in part on the perceived failure of Darwinian theory to explain design and purpose in biological systems. For a recent example, see *The Edge of Evolution* by Michael Behe (New York: Free Press, 2007).

Chapter 1. The goal of this brief outline of the conflict between materialistic and nonmaterialistic views of life, especially the nineteenth-century debate about vitalism, is merely to further the aims of *Life beyond Molecules and Genes'* narrative, not to provide the reader with an overview of older ideas about life's nature, or details of the debate over vitalism. If you would like to learn more about these subjects, start with Thomas S. Hall's broadly conceived and well-documented historical account: *History of General Physiology,* 2 vols. (Chicago: University of Chicago Press, 1969). Though it may seem incredibly useless in this day of molecules, reading a good translation of Aristotle's essential and formative ideas about life, especially his immensely influential masterpiece, the first biology text, *De Anima,* trans. Hugh Lawson-Tancred (London: Penguin Books, 1986), should be required reading for scientists and students of science alike. If one wants to learn more about how we got to modern biology, consider in addition Newton's *Principia* (1687, repr. Amherst, NY: Prometheus, 1995), Lavoisier's *Elements of Chemistry* (trans. Robert Kerr, Edinburgh, 1790, repr. Mineola, NY: Dover, 1965), and the original papers of Theodor Schwann, most importantly,

*Microscopical Researches into the Accordance in the Structure and Growth of Animals and Plants* (Bethesda, MD: Gryphon Editions, 2001).

As for Pasteur's experiments on the polarity of molecules, more than one hundred years after the discovery of dissymmetric forces, the chemical polarity of life remains a mystery. We have no idea of why nature chose the particular polarities it did.

Chapter 2. From a reductionist point of view, after cell theory, the protoplasmic theory of life seemed a logical next step, looking, as it did, further inward. To get a sense of its origins, see Thomas S. Hall's *History of General Physiology,* 2 vols. (Chicago: University of Chicago Press, 1969). Ultimately protoplasmic theory failed. See Louis V. Heilbrunn as he sounds a reluctant death knell in his 1952 text *An Outline of General Physiology* (Philadelphia: W. B. Saunders). In its place (in place of a physical explanation for life), biochemistry emerged. Its many accomplishments can be appreciated in any good biochemistry textbook. See, for example, David L. Nelson and Michael M. Cox's *Lehninger Principles of Biochemistry,* 4th ed. (New York: W. H. Freeman, 2004). For recent histories of biochemistry, look at *A History of Biochemistry* by Marcel Florkin (Amsterdam: Elsevier, 1990); *A Documentary History of Biochemistry* by Mikulas Teich and Dorothy M. Needham (Madison, NJ: Fairleigh Dickinson Press, 1992); and *Selected Topics in the History of Biochemistry* edited by G. Semenza, among others (Amsterdam: Elsevier Science, 1990).

Chapter 3. Modern ideas about life's origin have focused on "prebiosis," the chemical and physical events that preceded life. Though what has been done to illuminate these events is often intriguing, it is all unavoidably extremely speculative, and has little to say about the origin of the cell with its enclosing membrane, much less about life itself. Two scenarios are usually imagined for prebiotic development. In the first, molecular duplication (understood as the precursor to cellular reproduction) started the whole thing. Yet what molecules were first duplicated, how they did their autocatalytic

work, and under what environmental conditions remains somewhere between the very uncertain and totally obscure. In what is
probably the most popular notion today, called the "RNA world,"
RNA replication is seen as the first step on the path to life. See
Leslie Orgel's analysis in "The Origin of Life on Earth," *Scientific
American* 271, no. 4 (1994): 77–83. For an attempt to produce
autocatalytic RNA duplication in the laboratory, see W. K. John-
ston, et al., "RNA-Catalyzed RNA Polymerization: Accurate and
General RNA-Templated Primer Extension," *Science* 292 (2001):
1319–25. Though one gets the impression that the "reproduction
first" view is meant to implicate reproduction in life's origin, quite
the opposite is true. Inadvertently or not, what it proposes is that
molecular replication, "reproduction," did *not* embody first life,
but was its nonliving antecedent.

The alternative is that the advent of metabolic systems predated
molecular duplication. See Harold J. Morowitz, *Beginnings of
Cellular Life: Metabolism Recapitulates Biogenesis* (New Haven,
CT: Yale University Press, 1992). It has the same problem. In
averring that metabolic events preceded life, was its nonliving
antecedent, we are again left in the dark as to what transformation occurred from a nonliving chemical system to living cells.
For another view of life's chemical origins, see the recent *Life
Evolving* by Nobel Laureate Christian de Duve (Oxford: Oxford
University Press, 2002). Though much has been proposed about
the origin of life since A. I. Oparin's 1938 classic, *Origin of Life*
(for an English language translation, see Dover's 2003 edition),
even with its many problems it remains the source, the origin of
modern ideas about life's origin.

Chapter 4. To be sure, the biological materialist's view of life is
more complex and more nuanced than simply claiming an identity between life's material incarnation and life itself. This can be
seen in Daniel Koshland's list of life's causes ("The Seven Pillars
of Life," *Science* 295 [2002]: 2215–16), which provides a reasonable approximation of how I believe things are commonly
viewed among biologists today. Though no doubt many would
add reproduction to the list and choose to express things a bit

differently, I doubt that many would take great exception to what he has to say. As for the comments made about complexity in this chapter, they are discussed more fully in chapter 10, *Life as Complexity.*

Chapter 5. In this chapter we join the search for a sufficient property of life, focusing on the vascular perfusion of blood. To learn more about the anatomy and physiology of the circulatory system and the heart, see Arthur C. Guyton and John E. Hall, *Textbook of Medical Physiology,* 10th ed. (Philadelphia: W. B. Saunders, 2005), or Matthew N. Levy and Achilles J. Pappano, *Cardiovascular Physiology* (New York: Mosby, 2006).

As described, diffusion is key to understanding why the circulatory system with its heart and various conduits evolved, and it was all a matter of size. See J. B. S. Haldane, *On Being the Right Size* (Oxford: Oxford University Press, 1985). For a textbook-level discussion of diffusion from a biological perspective, see Meyer B. Jackson, *Molecular and Cellular Biophysics* (Cambridge: Cambridge University Press, 2006), or Roland Glaser, *Biophysics* (Berlin: Springer, 2004).

Chapter 6. Through the work and ideas of Claude Bernard, we learn of life's internal adaptations. To find out more about Bernard's work and his thinking, see *Claude Bernard and Animal Chemistry* by Fredric Lawrence Holmes (Cambridge, MA: Harvard University Press, 1974), as well as Bernard's own *Introduction to the Study of Experimental Medicine,* approx. 1865, trans. Henry Copley Greene (New Brunswick, NJ: Transaction Publishers, 1999), his *Lectures on the Phenomena of Life Common to Animals and Plants,* 1872–73, trans. Hebbel Hoff, Roger Guillemin, and Lucienne Guillemin (Springfield, IL: Charles C. Thomas, 1974), as well as his many research articles. For the concept of homeostasis, see Walter B. Cannon's *The Wisdom of the Body* (New York: W. W. Norton, 1932).

To learn about life's internal adaptations in any detail is a very large task. The only practical alternative to confronting an immense and extremely complex research literature covering essentially all of modern experimental biology are—despite their many categorical deficiencies—various textbooks. For example, Arthur C. Guyton and John E. Hall, *Textbook of Medical Physiology*, 10th ed. (Philadelphia: W. B. Saunders, 2005); Daniel L. Nelson and Michael M. Cox, *Lehninger Principles of Biochemistry* (New York: W. H. Freeman, 2004); Tristan G. Parlow, et al., *Medical Immunology*, 10th ed. (New York: McGraw-Hill, 2001); and Bruce Alberts, et al., *Molecular Biology of the Cell*, 4th ed. (New York: Garland Science, 2002).

Chapter 7. To learn about the lungs and respiratory system, see Michael P. Hlastala and Albert J. Berger, *Physiology of Respiration*, 2nd ed. (New York: Oxford University Press, 2001), or John B. West, *Respiratory Physiology: The Essentials*, 7th ed. (Philadelphia: Lippincott, Williams, Wilkins, 2004). For a comparative and evolutionary view of the respiratory system, see Kenneth Kardong, *Vertebrates: Comparative Anatomy, Function, Evolution*, 4th ed. (New York: McGraw-Hill Science, 2004), and D. T. Anderson, *Invertebrate Zoology*, 2nd ed. (New York: Oxford University Press, 2001). For the role of hemoglobin in presenting oxygen to our cells, see David E. Mohrman, *Cardiovascular Physiology*, 6th ed. (New York: McGraw-Hill, 2006), as well as Daniel L. Nelson and Michael M. Cox, *Lehninger Principles of Biochemistry* (New York: W. H. Freeman, 2004). To go deeper, the enormous experimental literature must be confronted. To give a sense of what is involved in doing so, one step in the process of providing oxygen to our cells and tissues— oxygen transport to tissue from blood—is the subject of a series of twenty-nine monographs, secondary sources published over more than a quarter century, the latest in 2008 (*Oxygen Transport to Tissue*, Vol. 29, ed. Kyung Kang, David K. Harrison, and Duane F. Bruley [Berlin: Springer]).

Chapter 8. As explained, it is popular today to imagine adaptations as being in DNA, as well as in other chemical or anatomical parts of the organism (see the Dawkins references above). This is the historical consequence of the neo-Darwinian synthesis that first wedded evolution to genetics. See R. A. Fisher, *The Genetical Theory of Natural Selection* (London: Oxford University Press, 1930), and J. B. S. Haldane, *The Causes of Evolution* (London: Longmans, Green, 1932). Though a grand scientific achievement and critical to the great advances in molecular biology that followed, reducing life to DNA and its protein genes led to discounting, or at least setting aside, a view that Darwin thought indispensable to his theory—that adaptations are propertic of whole individual organisms. See Darwin's *On the Origin of Species by Means of Natural Selection* (London: John Murray, 1859) or any of many reprints.

In a similar way, the claim of *Life beyond Molecules and Genes* that adaptations are what make certain material objects alive can only be true if they are phenomena of the whole living thing. It teaches that parts cannot provide for life's sufficiency. They are not alive themselves, nor can they in their own right make something living. It certainly should not be a radical point of view that life is a phenomenon of individual organisms, not their parts, though at times today it certainly seems so. The claim in this chapter that "life is not an autonomous, but a dependent phenomenon" is not meant to reject the idea that living things are "autonomous agents," as Stuart Kauffman calls organisms, but to merely point out that their actions are all reactions to the environment.

For an exploration of the concept of emergence as it applies to biology, see Harold J. Morowitz, *The Emergence of Everything: How the World Became Complex* (New York: Oxford University Press, 2004), and John H. Holland, *Emergence: From Chaos to Order* (Boston: Perseus Books, 1998).

Chapter 9. Although I have intentionally avoided chapter and verse, in chapter 9 we find ourselves in uncomfortable waters, confronting what is arguably Darwinism's most abiding controversy, the ubiquity of adaptations, or "adaptationism." Are all traits, all features of living things adaptations, or are they not? The danger that seemingly adaptationless traits pose to the theory of evolution is nothing less than its falsification. This is not a new realization, but dates back to the theory's beginnings, to Charles Darwin and Alfred Russel Wallace. See Charles Darwin's "On the Tendency of Species to Form Varieties; and on the Perpetuation of Varieties and Species by Natural Means of Selection," *Proceedings of the Linnean Society* 3 (1958): 45–62; *On the Origin of Species by Means of Natural Selection* (London: John Murray, 1859); *On the Various Contrivances by which British and Foreign Orchids Are Fertilized by Insects* (London: John Murray, 1862); and *The Descent of Man and Selection in Relation to Sex* (London: John Murray, 1871); and Alfred Russel Wallace's various writings—*Contributions to the Theory of Natural Selection* (New York: Macmillan, 1870), as well as a recent anthology edited by Andrew Berry, *Infinite Tropics* (London: Verso, 2002).

Today much that is controversial in science about the theory of evolution in science concerns this problem. For example, which *behavioral* traits are adaptive and which are not, and most importantly, are they what they are said to be (for example, and most prominently, is there such a thing as altruism)? There is also the task of attempting to specify the *origin and fate* of adaptive and nonadaptive traits, and grasping the glossary of terms that have been used to describe them—spandrels, preadaptation, devolution, cooptation, exaptation, co-opted adaptations, and so on. Finally, and perhaps most in dispute, there is the task of determining the material entities in which adaptations "reside." As suggested, contra Darwin, some evolutionary biologists today locate them in DNA, not whole organisms, while others see adaptive properties in populations of organisms. Sadly, some of these disputes, often simple definitional disagreements, have been hyperbolic, seemingly designed to promote controversy, infused with the superheated egos of some of the more notable participants. And this is not to mention the complaint of Intelligent Design devotees that natural selection leaves the evolution of complex mechanisms.

As discussed in chapter 11 and again here, much of the controversy derives from the difficulty of trying to ascertain which traits are adaptive and which are not. And this in turn is in great part the result of trying to decipher the obscure and impenetrable, unknown and undecipherable past in the present through the power of our inferences and imagination. For current-day phenomena, the difficulty lies in trying to distinguish what is adaptive from what is not and from what might be. It seems that almost any story we construct, any example of an adaptation we give, however reasonable and evocative, can be countered by an equally justified alternative explanation. It is on these rocky shores that the waves of the controversy continue to break.

Front and center in this hullabaloo are the attempts of scientists and philosophers to account for the personal and social behavior of humans as adaptations. Though long a part of Darwinian thinking, in recent times whole new fields of study—sociobiology, evolutionary psychology, and evolutionary anthropology—have emerged that attempt to navigate these difficult waters to learn nature's secrets past and present, often in what are unavoidably the most ambiguous of circumstances. However thoughtful the observer, however acute his or her intellect, the evolutionary antecedents of human behavior are of a sort with Darwin's peacock feathers, more often than not a matter of guessing, with little or no hope of knowing—fertile ground the scientist's imagination, but not much more—yielding what Stephen Gould, appropriating from Kipling, called "Just So Stories." Nonetheless, as explained, all is not lost if we limit ourselves to knowledge of the present and to express environmental circumstances. In this case, posited adaptive functions can be tested experimentally.

It is amid these difficulties and confusions that the viewpoint of *Life beyond Molecules and Genes* about life's source—like that of the theory of evolution by natural selection itself—rises or falls. It is critically dependent on the understanding that all traits, even if not adaptive in their own right, are associated with adaptations, or have been in the past. Otherwise, both evolution and life itself have causes that are not dependent upon adaptations, that are not dependent on natural selection. Though it may seem circular, if the claim that life can be defined by its adaptations depends on this understanding, then likewise the fact that we can indeed define life in this way and apparently no other offers support for the adaptationist's view of evolution.

To learn more about this controversy, consider the following sources. For the origin of the concept of the spandrel in biology, see S. J. Gould and R. C. Lewontin, "The Spandrels of San Marco and the Panglossian Paradigm: A Critique of the Adaptationist Programme," *Proceedings of the Royal Society of London* B 205 (1979): 581–98. Also see Stephen Jay Gould, "The Exaptive Excellence of Spandrels as a Term and Prototype," *Proceedings of the National Academy of Science* (USA) 94 (1997): 10750–55; R. C. Lewontin, *Biology as Ideology: The Doctrine of DNA* (New York: Harper Collins, 1993); Elliot Sober, "Six Sayings about Adaptationism," in *The Philosophy of Biology,* ed. David Hull and Michael Ruse (Oxford: Oxford University Press, 1998); and Stephen Gould's last book, his encyclopedic *The Structure of Evolutionary Theory* (Cambridge, MA: Harvard University Press, 2002).

For other views, see W. D. Hamilton, "The Evolution of Altruistic Behavior," *The American Naturalist* 97 (1963): 354–56; W. D. Hamilton, "The Genetical Evoluton of Social Behavior," *Journal of Theoretical Biology* 7 (1964): 1–52; George C. Williams, *Adaptation and Natural Selection* (Princeton, NJ: Princeton University Press, 1966); Helena Cronin, *The Ant and the Peacock: Altruism and Sexual Selection from Darwin to Today* (Cambridge: Cambridge University Press, 1992); U. Segerstrale, *Defenders of the Truth: The Battle for Science in the Sociobiology Debate and Beyond* (Oxford: Oxford University Press, 2000); and finally, Richard Dawkins' writings, especially his first book, *The Selfish Gene* (Oxford: Oxford University Press, 1976).

Chapter 10. Over the past ten to twenty years, there has been a great deal of interest in explaining life in terms of the ideas of complexity and self-assembly. Life—the most complex system known to us—is understood to be the consequence of self-assembly. In this view, complexity and self-assembly produce life. The writings of Stuart Kauffman are prominent in this area. See Stuart Kauffman in *Principles of Organization in Organisms: Proceedings of the Workshop on Principles of Organization in Organisms,* ed. Jay E. Mittenthal and Arthur B. Baskin (Boston: Perseus Publishing, 1992); Stuart Kauffman, *The Origins of Order: Self-Organization and Selection in Evolution* (New York: Oxford University Press, 1993); and Stuart Kauffman, *At Home in the Universe: The Search for the Laws of Self-Organization and Complexity* (New York: Oxford University Press, 1995). More recently, Kauffman has defined life in different terms, as a "physical system capable of self-reproduction and also capable of performing at least one thermodynamic work cycle" ("What Is Life?" in *The Next Fifty Years,* ed. John Brockman [New York: Vintage Books, 2002], 126–41). But as with complexity, neither reproduction nor work is a sufficient property of life.

For the mathematical roots of the notion of "complexity numbers," see the writing of Gregory J. Chaitin. For example, Gregory J. Chaitin, *Exploring Randomness* (Berlin: Springer, 2001), and Gregory J. Chaitin and Paul Davies, *Thinking about Godel and Turing: Essays on Complexity, 1970–2007* (Hackensack, NJ: World Scientific Publishing Company, 2007).

For other perspectives, see Hubert P. Yockey, *Information Theory, Evolution and the Origin of Life* (Cambridge: Cambridge University Press, 2005); Niels Henrik Gregerson, ed., *From Complexity to Life: On the Emergence of Life and Meaning* (New York: Oxford University Press, 2003)—see the chapter by Gregory J. Chaitin on complexity; Bernd-Olaf Kuppers' *Information Theory and the Origin of Life* (Cambridge, MA: MIT Press, 1990); Bruce H. Weber, David J. Depew, and James D. Smith, eds., *Entropy, Information and Evolution: New Perspective on Physical and Biological Evolution* (Cambridge, MA: MIT Press, 1988); as well as Harold J. Morowitz's *The Emergence of Everything: How the World Became Complex* (New York: Oxford University Press, 2004), and John H. Holland's *Emergence: From Chaos to Order* (Boston: Perseus Books, 1998).

Chapter 11. In the end, it is not life's physical and chemical embodiment, its complexity, information, organization, metabolic cycles (see Harold J. Morowitz, *Beginnings of Cellular Life: Metabolism Recapitulates Biogenesis* [New Haven, CT: Yale University Press, 1992]), nor viewing life as a system (see Robert Rosen, *Life Itself* [New York: Columbia University Press, 1991]) that provides us with knowledge of its sufficient qualities. Neither singly nor together do they allow us to distinguish the living from the nonliving. As such, they cannot provide us with an adequate definition of life. After all is said and done, life's sufficient properties are to be found in its adaptations and in them alone.

Though the idea of biological adaptations is ancient, to learn about our modern understanding we must of course begin with Charles Darwin and Alfred Russel Wallace, who joined the notion of adaptations to natural selection (see above for citations). In *On the Origin of Species,* Darwin provides many thoughtfully chosen examples of adaptations drawn from a wide range of circumstances and species. To understand how our understanding has evolved since Darwin's time, see Theodosius Dobzhansky, *Genetics and the Origin of the Species* (New York: Columbia University Press, 1937); Ernst Mayr, *Systematics and the Origin of the Species* (New York: Columbia University Press, 1942), as well as his last book for a general audience, *This Is Biology* (Cambridge, MA: Harvard University Press, 1997); George C. Williams, *Adaptation and Natural Selection* (Princeton, NJ: Princeton University Press, 1966); Verne Grant, *The Origin of Adaptations* (New York: Columbia University Press, 1963); Marjorie Grene, ed., *Dimensions of Darwinism: Themes and Counterthemes in Twentieth-Century Evolutionary Theory* (Cambridge: Cambridge University Press, 1983); and Niles Eldredge's *Unfinished Synthesis: Biological Hierarchies and Modern Evolutionary Thought* (New York: Oxford University Press, 1985) and *Fossils* (Princeton, NJ: Princeton University Press, 1991). For recent technical monographs, see M. R. Rose and G. V. Lauder, ed., *Adaptation* (San Diego: Academic Press, 1996), and Steven Hecht Orzack and Elliot Sober, eds., *Adaptationism and Optimality* (Cambridge:

Cambridge University Press, 2001). There are also the many engaging books and articles by the late Stephen Jay Gould. For example, *Ontogeny and Phylogeny* (Cambridge, MA: Harvard University Press, 1977); *Ever Since Darwin* (New York: W. W. Norton, 1977); *Hen's Teeth and Horse's Toes* (New York: W. W. Norton, 1983); *An Urchin in the Storm* (New York: W. W. Norton, 1987); *Wonderful Life* (New York: W. W. Norton, 1989); and to get a sense of the scope of the enterprise since Darwin, dip into Gould's *The Structure of Evolutionary Theory* (Cambridge, MA: Harvard University Press, 2002).

In chapter 9, I give music as an example of an adaptation. In a recent book, *The Singing Neanderthals* (Cambridge, MA: Harvard University Press, 2005), Steven Mithen discusses this at length. Whether the particular situation I described in chapter 9 about the experience of concertgoers is convincing or not, music provides a good example of the difficulty of assessing whether a phenomenon is adaptive without carrying out experiments to make that determination. We can claim, as my telling of the concertgoers' story does, that listening to music is adaptive for humans simply because I have personally experienced its benefit—that is, it is adaptive subjectively. This said, Darwin thought that music posed a significant difficulty for his theory. It was not clear to him that it was an adaptive property. Eventually he satisfied himself that along with the feathers of peacocks and the elaborate and varied color and form of flowers, music (the bird's song) has sex appeal (of course, he had not heard me sing). See Darwin's *The Descent of Man and Selection in Relation to Sex*, 2 vols. [London: Murray, 1871]). The difficulty music presents is, like so many human activities, that it does not generalize well. In the example I gave, it may reduce stress for me, while producing it in someone else *at the same time, under the same environmental circumstances*. And how do we explain martial music designed to give soldiers courage as they prepare for and enter battle to give up their lives? It certainly does not seem to work in the service of the soldier's survival, but rather causes him to slouch toward his death. Is this counteradaptive music? What is science to make of such phenomena—phenomena from which we cannot generalize from specific examples, specific instances? With a nod to quantum theory, what are we to make of things and events that can be both one way and its opposite, adaptive and counteradaptive?

Conclusion. To read Dawkins' evocative fantasy, see "Son of Moore's Law," in *The Next Fifty Years,* ed. John Brockman (New York: Vintage Books, 2002), 145–58, and *Unweaving the Rainbow* (New York: Houghton Mifflin, 1998).

A professor of anthropology and genetics at Pennsylvania State University, Mark Shriver, is attempting to realize Dawkins' fantasy of genetically determining facial appearance as a forensic alternative to police artist sketches (Gautam Naik, *Wall Street Journal,* March 27, 2009). Professor Shriver hopes to statistically correlate an individual's genetic makeup with facial features such as the shape and dimensions of the lips and nose. For instance, genetic markers related to ethnic ancestry should be correlated to facial appearance. Broader noses are more common among Africans, stubby ones among Germans and Scandinavians, and curved ones among Jews and Arabs. However, it is not clear that images rooted in group rather than individual differences would offer an improvement over a police artist's sketch drawn in accordance with a witness' description of a particular person. At any rate, correlating genes of whatever sort of appearance is not the same as showing that particular genes *produce*, no less embody, features of a particular shape and size. For a related study see Y. C. Klimentidis and M. D. Shriver, "Estimating Genetic Ancestry Proportions from Faces," *PLoS ONE* 4 (2009): 4460.

Reflection. There is an enormous literature on the relationship between religion and science, especially if we include the dichotomy of reason and belief, spanning more than two millennia. The references that follow are primarily concerned with the Hebrew Bible, Greek philosophy, and Christian scripture, not Buddhism, Islam, Hinduism, Confucianism, and numerous other religious traditions. For someone who is not particularly religious, but considers him or herself at least culturally part of the western religious tradition, an open-minded reading of Genesis is a critical starting point. Many early religious thinkers thought about the problem of Genesis and of biblical interpretation more generally, including Philo (Kenneth Schenck, *A Brief Guide to Philo* [Louisville, KY: Westminster John Knox Press, 2005]; Philo, *The Biblical Antiquities of Philo*, trans. M. R. James [Whitefish, MT: Kessinger Publishing, 2004]), St. Augustine (St. Augustine, *Confessions*, trans. Henry Chadwick [Oxford: Oxford University Press, 1998]; Vernon J. Bourke, ed., *St. Augustine: City of God* trans. Marcus Dods [Peabody, MA: Hendrickson, 2009]), and Maimonides (Moses Maimonides, *The Guide to the Perplexed*, 2 vols., trans. Shlomo Pines [Chicago: University of Chicago Press, 1963]. In post-Enlightenment times, theologians from all sorts of religious perspectives have written about the relation between science and religion. For example, for Catholics, see James J. Walsh, *The Popes and Science: The History of the Papal Relations to Science during the Middle Ages and Down to Our Own Time* (1908; repr., Whitefish, MT: Kessinger Publishing, 2003); Pierre Teilhard de Chardin, *Le Milieu Divin* (New York: Harper Perennial, 2001); Joseph Ratzinger, *Christianity and the Crisis of Cultures* (Fort Collins, CO: Ignatius Press, 2006). For Protestants, see Arthur Peacocke, *Theology for a Scientific Age: Being and Becoming—Natural, Divine and Human* (Minneapolis, MN: Augsburg Fortress Publishers, 1993), Philip Clayton, ed., *All That Is: A Naturalistic Faith for the 21st Century* (Minneapolis, MN: Fortress Press, 2007); Philip Clayton and Arthur Peacocke, *In Whom We Live and Move and Have Our Being: Panentheistic Reflections on God's Presence in a Scientific World* (Grand Rapids, MI: Eerdmans, 2004); John Polkinghorne, *Science and Theology* (Minneapolis, MN: Fortress Press, 1998); Peter Kreeft and Ronald Tacelli, "Twenty Arguments for the Existence of God," in *Handbook of Christian Apologetics* (Westmont, IL: InterVarsity Press, 1994). For Jews, see Geoffrey Cantor and Marc Swetlitz, eds., *Jewish Tradition and the Challenge of Darwinism* (Chicago:

University of Chicago Press, 2006); Hans Jonas, *The Phenomenon of Life* (New York: Harper and Row, 1966); and Natan Slifkin, *The Challenge of Creation* (New York: Yashar Books, 2006).

As for scientists, Albert Einstein's thoughts are probably the most often quoted. See Max Jammer, *Einstein and Religion* (Princeton, NJ: Princeton University Press, 1999); Robert N. Goldman, *Einstein's God: Albert Einstein's Quest as a Scientist and as a Jew to Replace a Forsaken God* (Lanham, MD: Rowman and Littlefield 1995); Albert Einstein, *Ideas and Opinions* (New York: Modern Library, 1954). In addition, recent books by scientists include Francis Collins' *The Language of God: A Scientist Presents Evidence for Belief* (New York: Free Press, 2006), and Stephen Jay Gould's *Rocks of Ages: Science and Religion in the Fullness of Life* (New York: Ballantine Books, 1999).

Three recent and very popular books by atheists are Daniel Dennett's *Breaking the Spell: Religion as a Natural Phenomenon* (New York: Penguin, 2007); Christopher Hitchens' *The Portable Atheist: Essential Readings for the Nonbeliever* (Cambridge, MA: De Capo Press, 2007); and Richard Dawkins' *The God Delusion* (London: Bantam Press, 2006). For the agnostic perspective, see Bertrand Russell, *Am I an Atheist or an Agnostic?* (Pittsburg, KS: E. Haldeman-Julius, 1949); Bertrand Russell, *Why I Am Not a Christian and Other Essays on Religion and Related Subjects,* ed. Paul Edwards (New York: Touchstone, 1967); Thomas Henry Huxley, *Man's Place in Nature: Man's Place in Nature and Other Anthropological Essays* (Whitefish, MT: Kessinger Press, 2005), as well as David Hume's *Dialogues Concerning Natural Religion,* ed. Richard Popkin (Indianapolis, IN: Hackett Publishing, 1998). For the philosophical and religious skeptic, there is Panayot Butchvarov, *Skepticism about the External World* (New York: Oxford University Press, 1998).

Other recent books on the relationship between science and religion are Karl Giberson, *Worlds Apart: The Unholy War between Religion and Science* (Kansas City, MO: Beacon Hill Press, 1993); Gary Ferngren, ed., *Science and Religion: A Historical*

*Introduction* (Baltimore, MD: Johns Hopkins Press, 2002); Paul Kurtz, Barry Karr, and Ranit Sandhu, *Science and Religion: Are They Compatible?* (Buffalo, NY: Prometheus Books, 2003); Ken Wilber, *The Marriage of Sense and Soul: Integrating Science and Religion* (New York: Broadway, 1999); and Kenneth R. Miller, *Finding Darwin's God: A Scientist's Search for Common Ground Between God and Evolution* (New York: Harper Perennial, 2007).

On the current Intelligent Design debate, see Michael Behe, *The Edge of Evolution* (New York: Free Press, 2007); William Dembski and Michael Ruse, eds., *Debating Darwin: From Darwin to DNA* (Cambridge: Cambridge University Press, 2007); Jonathan Wells, *The Politically Incorrect Guide to Darwinism and Intelligent Design* (New York: Regnery Publishing, 2006); Michael Ruse, *Darwin and Design: Does Evolution Have a Purpose?* (Cambridge, MA: Harvard University Press, 2004); John Wilson and William A. Dembski, *Uncommon Dissent: Intellectuals Who Find Darwinism Unconvincing* (Wilmington, DE: ISI Books, 2004); Michael Denton, *Icons of Evolution: Science or Myth?* (New York: Regnery Publishing, 2002); David Stove, *Darwinian Fairytales* (New York: Encounter Books, 1995); Phillip E. Johnson, *Darwin on Trial* (Downers Grove, IL: InterVarsity Press, 1993); Michael Denton, *Evolution: A Theory in Crisis*, 3rd ed. (Bethesda, MD: Adler and Adler, 1986). Finally, for a discussion of scientists as gods, see David Berlinski, "The God of the Gaps," *Commentary* magazine, April 2008, 34–40.

# Index

acidity, 86

action: adaptations in, 111; complexity of, 147–48; reaction and, of living things, 118–19

adaptability, 42; Darwin on, 110, 157; as guarantee of aliveness, 44; limbs of animals as measure of, 167–68; Richet on, 68; sufficient properties relation to, 67

adaptation(s): abundance of internal, 70–71, 79–80; Bernard discovering, 79–80; biological, as basis for aliveness, 9–10, 69, 104–6; Cannon on, of frogs, 155; complexity and, 153–54; as consequences of evolution, 156; in context/action, 111; as contextual/contingent, 156; as contingent rather than inherent, 104; as emergent from material features, 116; Darwinian, 100; Darwinian v. Bernardian, 89–91; death and, 120; devolved, 129; environmental conditions and, 101, 112, 156; exterofective v. interofective, 90; insufficiency of material elements, 116; internal, 10; internal v. external, 89–91; life/aliveness as dependent on, 9–10, 104, 106, 156; material features in context as, 118; as material incarnations, 107; material incarnations, evolution and, 112–13; measurability of mechanisms underlying, 169; natural selection and, 130; as sufficient properties of life, 156; as transcendant/emergent, 156; of vascular perfusion, 114

adaptive capacity, 164–66

adaptive properties, 107, 111, 154, 156, 165; of chin, 126; of inanimate objects, 116–17; material elements and, 116; of religion, 128; of rock/symphony concerts, 128; of soft spandrels, 126–27; sufficient properties of life revealed by its, 155. *See also* nonadaptive properties.

adenosine tri-phosphate (ATP), 60

agility, 165

alveoli, 97

aneurysm, 53

*anima*, 102

animal movement, 14

animals: Dawkins on computing, from genomes, 173; humans as information-producing, 132; limbs of, as measurement of adaptability, 167–68

animate beings, 31–32

appendages, 61

architecture, 123–24

Aristotle, 14, 37, 102, 131

arteries, 65

arterioles, 75

atherosclerosis, 53